# Hero and Wanderer

## A Reappraisal of Pluto's Astrological Significance

By Jim Trader

**Hero and Wanderer**

Fifth Estate, Post Office Box 116,
Blountsville, AL 35031

First Edition
Cover Designed by An Quigley

Printed on acid-free paper

Library of Congress Control No: 2011942936
ISBN: 9781936533190

Fifth Estate, 2011

**Jim Trader**
**Natal Chart**
Feb 4 1976, Wed
0:08 am CST +6:00
Milwaukee, Wisconsin
43°N02'20" 087°W54'23"
*Geocentric*
*Tropical*
*Koch*
*Mean Node*

# Table of Contents

**How To Use This Book**................................................**11**

**Chapter 1. Pluto's General Significance**.........**13**
Pluto: Planet or Not?.....................................................**13**
A Strange Complacency....................................**14**
A Window of Opportunity....................................**15**
*Table of Pluto in the Signs*.....................................16
New Observations of Pluto.....................................**17**
Terms and Definitions..........................................**18**
The Hero's Journey...............................................**20**

**Chapter 2. Pluto's Individual Influence**.........**23**
Pluto in the Signs and Houses.................................**23**
*Table of Sign Themes*...........................................24
*Table of House Themes*.........................................24
    **Pluto in Cancer**.................................................**25**
    **Pluto in Leo**......................................................**26**
        1st House...............................................27
        2nd House...............................................27
        3rd House...............................................27
        4th House...............................................28
        5th House...............................................28
        6th House...............................................28
        7th House...............................................29
        8th House...............................................29
        9th House...............................................29
        10th House...............................................30
        11th House...............................................30
        12th House...............................................30
    **Pluto in Virgo**...................................................**31**
        1st House...............................................32
        2nd House...............................................32

3rd House.............................................32
4th House.............................................33
5th House.............................................33
6th House.............................................33
7th House.............................................34
8th House.............................................34
9th House.............................................34
10th House............................................35
11th House............................................35
12th House............................................35
**Pluto in Libra.....................................36**
1st House.............................................36
2nd House.............................................37
3rd House.............................................37
4th House.............................................37
5th House.............................................38
6th House.............................................38
7th House.............................................38
8th House.............................................39
9th House.............................................39
10th House............................................40
11th House............................................40
12th House............................................40
**Pluto in Scorpio..................................41**
**Pluto in Aspect to other Planets..........42**
*Table of Planet Themes*............................43
*Table of Aspect Themes*............................43
**Pluto-Sun Aspects...............................44**
Conjunction.........................................45
Sextile.............................................46
Square..............................................46
Trine...............................................46
Quincunx............................................47
Opposition..........................................47
**Pluto-Moon Aspects.............................47**
Conjunction.........................................49
Sextile.............................................49
Square..............................................50
Trine...............................................50

Quincunx............................................51
Opposition..........................................51
**Pluto-Mercury Aspects**...............................**52**
Conjunction.........................................52
Sextile.............................................53
Square..............................................53
Trine...............................................54
Quincunx............................................54
Opposition..........................................54
**Pluto-Venus Aspects**...................................**55**
Conjunction.........................................55
Sextile.............................................56
Square..............................................56
Trine...............................................57
Quincunx............................................57
Opposition..........................................57
**Pluto-Mars Aspects**...................................**58**
**Pluto-Mars Conjunction, Sextile, and Trine**.........**59**
**Pluto-Mars Square, Quincunx and Opposition**.....**60**
**Pluto-Jupiter Aspects**...............................**60**
Conjunction.........................................61
Sextile.............................................62
Square..............................................62
Trine...............................................63
Quincunx............................................63
Opposition..........................................64
**Pluto-Saturn Aspects**...............................**64**
Conjunction.........................................65
Sextile.............................................65
Square..............................................66
Trine...............................................66
Quincunx............................................67
Opposition..........................................67
**Pluto-Uranus Aspects**...............................**67**
**Pluto-Neptune Aspects**...............................**69**
**Final Comments on Pluto's Individual Influence**...........**70**

## Chapter 3. Pluto's Generational Influence..........74
An analysis of Pluto in Cancer's Hero's Journey............74
The 7-Stage Life-Cycle of a Pluto Generation.................76
    Example: Pluto in Cancer.................................78
Other Pluto Generational Effects...............................80

## Chapter 4. Pluto's Transiting Influence..............83
    Pluto in the Houses.................................84
    1$^{st}$ House..............................................85
    2$^{nd}$ House..............................................85
    3$^{rd}$ House..............................................86
    4$^{th}$ House..............................................87
    5$^{th}$ House..............................................88
    6$^{th}$ House..............................................89
    7$^{th}$ House..............................................89
    8$^{th}$ House..............................................90
    9$^{th}$ House..............................................91
    10$^{th}$ House.............................................91
    11$^{th}$ House.............................................92
    12$^{th}$ House.............................................93
Aspects From Transiting Pluto to Natal Planets...............93

## Chapter 5. Pluto's Synastric Influence...............96

## Afterword.................................................98
    A Final Note.................................................101

## Appendix: Statistical Data............................103
    *Table of Baseline Group Data*............................104
    *Tables of Astrologer Group Data*........................107
    *Table of Businesspeople Group Data*....................108
    *Tables of Entertainment Group Data*....................109
    *Tables of Infamous Group Data*.........................110
    *Table of Literature Group Data*.........................111
    *Tables of Politics Group Data*..........................112
    *Tables of Royalty Group Data*..........................113
    *Tables of Science Group Data*..........................114

*Tables of Sports Group Data*..............................115
*Tables of Visual Arts Group Data*.......................117
*Summary Table of Pluto Aspect Profiles*...............118

## How To Use This Book

For anyone interested in astrology and its impact on life here on Earth, I like to think this book is essential reading. However, the fact is that it is largely written for an astrologically-astute audience, and therefore assumes that the reader is already familiar with astrological terms such as Sign, House and Aspect. This book also assumes that the reader understands the difference between a natal and transiting astrological influence, and that he or she knows what "synastry" is. If you have heard this astrological jargon before but aren't sure exactly what it means (or if you are familiar with some of the terms but not all of them), there is a good chance all the Chapters of this book will still prove useful for you, as any terms you may not be immediately familiar with can be understood through context. Readers who are totally unfamiliar with astrological jargon may be best served by sticking to the General Significance Chapter, and perusing the Afterward to provide further food for thought.

Like many astrology texts, this book is a combination of reference work and bully pulpit. Specific information for interpreting Pluto's influence in the natal chart and in world affairs can be found in the Individual Influence, Generational Influence, Transiting Influence and Synastric Influence Chapters. The Table of Contents is meant specifically to direct the reader to this particular kind of information, if that is what the reader is seeking. For those who are more interested in a broader understanding of Pluto's astrological significance, the General Significance Chapter and the Afterward may prove rewarding. For those interested in the methodology used to arrive at these conclusions, and how Pluto seems to influence broad groups of people with similar interests and traits, the Appendix showing the statistical data and methodology used may prove helpful.

All caveats aside, this book is meant to be both enjoyable and enlightening for anyone who picks it up.

# Chapter 1: General Significance

Death and transformation: This is as good a summary as any of how most of the astrological community understands Pluto's astrological influence in the early 21$^{st}$ century. Although observation reveals that this Planet's earthly correspondences are far more nuanced and nurturing than this vaguely frightening description might suggest, the subject of Pluto is certainly one that intimidates many in the astrological field. Both astronomically and astrologically, Pluto embodies a dark and mysterious place we find somehow disruptive even from a great distance, a place whose nature and influences most of us would feel most comfortable avoiding. However, like puberty or political discontent, Pluto's influence is unavoidable and is at its most dangerous when ignored. The Planet Pluto represents those unknowns that we must eventually experience and come to terms with, and which act as doorways from one state of being to another. When our fears of the unknown and of change are overcome and we allow the light of consciousness to shine on those things represented by this Planet, Pluto's influence becomes like a mirror that reflects ultra-violet and x-ray radiation back at us rather than the normal visible-spectrum light we are used to. The x-ray images we then receive show us treasures, shadings, structures and fault-lines integral to who we are that no other source can reveal. Acting on the knowledge Pluto provides us opens the door to a maturity and to powers that would otherwise be unavailable to us. Vengeance and manipulation are two negative traits often associated with Pluto's influence, but these only come to the fore when we try to avoid Plutonic influence out of fear. Strange, then, that we've collectively chosen to represent this Planet with a name—Pluto, a greatly feared god of the underworld—that encourages its negative traits while minimizing the positive.

## Pluto: Planet or Not?

When observed without mythological baggage, the Planet Pluto clearly embodies an exception that precipitates crisis. This is evident even in Pluto's astronomical traits, which fuel an ongoing debate among astronomers as to whether Pluto is a "planet," a "dwarf planet," or something else entirely. Pluto's small size relative to other

Planets arguably demotes it from Planetary status, as does the fact that there are apparently other large objects sharing Pluto's orbit. According to astronomers, to be considered a Planet, a celestial body must "clean its orbit," meaning that it must sweep anything else of reasonable size out of its path or capture these foreign objects in its own gravity well as it proceeds in its orbit around the Sun. Pluto fails to clean its orbit. And yet, despite these glaring deficiencies in the eyes of many astronomers, Pluto's place as a "Planet" appears to have remained unshaken in many people's minds, including the minds of most astrologers and even a minority of astronomers. Pluto's size (small for a Planet, but large for just about anything else), its independent solar orbit and its possession of a moon all reinforce the perception of Pluto as a Planet in many people's minds.

What I believe to be the real (and rarely discussed) crux of the debate is that Pluto wanders from the plane of the ecliptic, the abstract plane in space most Planets' orbits follow. This comfortable convention allows the solar system to be imagined as a disc rather than a sphere in space. For second-class citizens of the solar system like comets, asteroids or moons, wandering from the ecliptic is acceptable behavior. However, the fact that all the Planets—the most astronomically significant bodies in the Solar System other than the Sun—keep to the ecliptic is what defines the convention of the ecliptic. Giving a celestial body that does not hold to this convention Planetary status is a threat to the model of the solar system astronomers have become comfortable with over millennia. And yet there Pluto is, claiming a place as a member of the Planetary fraternity in the minds of many of us here on Earth even as it defies the conventions astronomers have set for this fraternity.

**A Strange Complacency**

A different facet of Pluto's influence is visible when one focuses on its impact on the astrological community. Discovered in 1930 and taking about 248 years to make one full circuit of the Zodiac, the mere 80 years astrologers have had to correlate Pluto's motion to circumstances here on Earth is unlikely to have revealed every facet of its influence yet. In the brief time astrologers have had to observe it, however, four facts regarding this celestial body have become quite clear. The first is that it is definitely a Planet, in that its correspondences to phenomena here on Earth are of an active nature (constant and independent of its relation to other Planets) rather than passive nature. Secondly, its correspondences involve the phenomena of power, disruption, and the unseen in various ways. Thirdly, Pluto affects large groups of people in ways that Planets closer to the Sun (from Saturn inward) do not. The fourth fact is the strangest: Very few astrologers seem inclined to look any more deeply than these three facts into what Pluto symbolizes, appearing to be quite satisfied with this very limited information.

This lack of curiosity is very atypical behavior. Astrological magazines and the shelves of metaphysical bookstores are full to bursting with the careful research and wild speculation astrologers lavish on their profession, giving testament to their penetrating minds and their powerful curiosity regarding most astrological subjects. Pluto appears to be a rare exception to this rule. While logic suggests (and initial observation supports) that ongoing observation will refine and expand our understanding of Pluto's correspondences here on Earth, yielding unexpected and probably useful results, astrologers by and large appear strangely complacent regarding Pluto. Discussion of Pluto's significance among astrologers and in astrological texts is usually limited to correlations to events or phenomena that occurred around the time Pluto was discovered, or to the mythology surrounding gods of death, destruction and the underworld. It seems premature to conclusively define Pluto's effects and symbolism like this, after so relatively little observation. I am very interested to see how the astrological community reacts to my observations, both of Pluto and of the community, since any reactions may provide me with further clues as to why this is so.

## A Window of Opportunity

This book has been written in the midst of one of the best possible times to gather observational data on Pluto's correspondences to people and events here on Earth. The period of time from the mid-1990s through the early 2020's may be the most revealing part of Pluto's current 248-year astrological cycle. As the table below shows, Pluto began moving (relatively) quickly through the Zodiac in the late 1950's and will continue to do so through the early 2020's. This means that a larger than normal number of Pluto generations is alive and active around the time this book was written.

15

| Sign | Dates |
|------|-------|
| Pluto in Aries | Apr 1822-Jan 1853 (31 yrs) |
| Pluto in Taurus | Apr 1852-Jul 1882 (30 yrs) |
| Pluto in Gemini | Jul 1882-May 1914 (32 yrs) |
| Pluto in Cancer | Jul 1913-Jun 1939 (26 yrs) |
| Pluto in Leo | Oct 1937-Jun 1958 (21 yrs) |
| Pluto in Virgo | Aug 1957-Jul 1972 (15 yrs) |
| Pluto in Libra | Oct 1971-Aug 1984 (13 yrs) |
| Pluto in Scorpio | Nov 1983-Nov 1995 (12 yrs) |
| Pluto in Sagittarius | Jan 1995-Nov 2008 (13 yrs) |
| Pluto in Capricorn | Jan 2008-Nov 2024 (16 yrs) |
| Pluto in Aquarius | Mar 2023-Jan 2044 (21 yrs) |
| Pluto in Pisces | Mar 2043-Feb 2068 (25 yrs) |

Astute readers may note that in the table above there is some overlap between the end of Pluto's tenure in one Sign and its entry into the another, and also that the total time given in the table for Pluto's cycle through the Zodiac comes to 255 years, while I stated previously that it has an orbit of 248 Earth years in length. Pluto's frequent retrograde motion is the reason for these ambiguities. Since Pluto often passes over the same degree of the Zodiac multiple times during a twelve month period, there is some room for interpretation regarding when Pluto "definitely" leaves one Sign and enters another when discussing its astrological influence, or for when it has "definitely" returned to the same degree of the Zodiac.

However debatable the exact time of Pluto's transition from one Sign to another may be, it's clear that after Pluto's current sojourn in the Sign of Capricorn the length of Pluto's stay in various Signs will once again lengthen significantly. Assuming Pluto spends about the same amount of time in each Sign each time it circles the Zodiac, the next time we will get an adult population with the diversity of natal placements in Pluto we had in 2010 will be sometime around 2265. The last time we had this much diversity in examples of Pluto's natal influence was in 1755, but since the astrological community was unaware of Pluto at the time, studying its correspondences was not possible.

All in all, it seems that this is a better time than most to study and document Pluto's astrological influence, for the sake of drawing more complete conclusions later.

## New Observations Of Pluto

I have been noting Pluto's correspondences in my own life and in the lives of those around me for several years, and I have observed some apparently significant correlations I have not heard or read any other astrologers to comment upon. The main purpose of this book is to point out these correlations, show how they have proven useful in and of themselves, and encourage others to use these observations as a springboard toward further refinement of our Plutonic understanding before the current window of observational opportunity closes.

Although the contents of this book certainly constitute an expansion of the astrological discourse regarding this Planet's symbolism, they can't be taken as the final word. We'll probably have to wait until 2180 or so—when Pluto has returned to the same place in the Zodiac it was when it was discovered—before we can start drawing definitive conclusions regarding this Planet's complex and fascinating influence. It is my hope that those who read this book will take advantage of the opportunity between now and 2023, to watch Pluto's correspondences more carefully and learn as much about it as possible.

Given the general inertia I've observed in the astrological community regarding researching and interpreting Pluto's influence, I applaud 20th Century astrologers such as Donna Cunningham, Steven Forrest, Jeff Green, Dane Rudhyar and Howard Sasportas for swimming against the tide and bringing us their original analyses of Pluto's symbolism. They have each taken a pocketknife to the sleeves of this straight-jacketed discourse and so allowed some circulation back into the limbs of the study of Pluto's influence. However, the limited freedom of movement these efforts have provided have not unlocked the padded cell of prejudice and timidity that still confines the body of this area of study. To my ears, Pluto's howls for recognition still seem deadened by the soundproofing of premature tradition. Hopefully this book will provide a key that unlocks this room, so Pluto's frightening howls can be replaced by reasoned dialog

17

and unfettered observation concerning the effects it has on us both collectively and individually.

Perhaps Pluto itself is influencing our discourse on its symbolism, creating the very tendencies that make discussions of it so limited and so fraught. Looking at the charts of twelve prominent astrologers active in the 20th century (those of Alice Bailey, Donna Cunningham, Jeanne Dixon, Steven Forrest, Linda Goodman, Jeff Green, Robert Hand, Karl Ernst Krafft, Dane Rudhyar, Howard Sasportas, Richard Tarnas, and Noel Tyl), I find that seven have an Aspect from Pluto to their Moon. As I point out later in this book, the energies of the Moon and Pluto often do not harmonize well; Pluto brings a compulsive darkness to the landscape of the unconscious represented by the Moon, often making this landscape the home of fearsome monsters that the native can neither understand or resist. It seems possible to me that astrologers with a Pluto-Moon Aspect (and there appear to have been a lot more of them in the 20th century than has been the norm throughout history) perceive Pluto as symbolizing their inner demons. If this is true, it may be a contributing factor to why discussion of Pluto is so limited, and especially why many astrologers avoid directly associating Pluto and the individual personality. This disassociation is most often accomplished by labeling Pluto as a "generational" or "karmic" influence rather than a personal one. Interestingly, the five 20th century astrologers I've observed to take the most constructive, innovative and reasoned approaches to Pluto's influence—Donna Cunningham, Steve Forrest, Jeff Green, Dane Rudhyar, and Howard Sasportas—are the same astrologers whose natal charts lack a Pluto-Moon Aspect. The absence of Pluto's influence is often just as meaningful its presence. I go into this in great detail in the Personal Influence Chapter and the Statistical Data Appendix, but this serves very well as an initial example.

**Terms and Definitions**

Before delving further into the specifics of what I've observed of Pluto's astrological significance, a few important terms I'm going to use need to be defined. The first is the term "power". In a Plutonic context, power is the ability of an individual or group to force the world (circumstances and others) to adapt to them. Where this power stems from, where it flows, how it is used and the consequences of its use all correlate strongly to Pluto, and so I believe the term "power" must be understood clearly if we are to correctly interpret anything else about Pluto's astrological meaning.

The second is the term "integrity," by which I mean the ability to match actions to stated principles. An example of the presence of integrity would be that of someone who claims to value honesty consistently sacrificing advantage or gain to tell the truth. An example of the absence of integrity might be that of someone who claims to follow an uncompromisingly pacifistic faith such as Buddhism or Christianity who chooses the career of a soldier, thereby basing his or her life, reputation and fortune on their ability to kill others and destroy property, and on supporting others to kill and destroy. Such a person fails to match their actions to their stated principle. Integrity is important when discussing Pluto's influence because it appears to be the defining factor in whether the power represented by Pluto brings ultimate success or failure. Integrity appears to be the only thing that ensures Plutonic power is used properly. Its absence appears to guarantee that whatever short-term benefits power brings, ultimately a Pyrrhic victory is the best its wielder can hope for. Examples of how this manifests are included in the Individual Influence and Generational Influence chapters.

Power being the essence of Pluto's influence, Pluto is the key to describing both empowerment and crisis in our individual and collective lives. "Empowerment" and "crisis" are the two other terms that require definition before my description of Plutonic observations can continue. I define empowerment as the process by which one maintains or increases one's power, and I define crisis as a situation that threatens one's goals, health or integrity. Before illustrations of these can be provided, a few more of the broad strokes of Pluto's influence have to be described.

Not only does Pluto appear to represent empowerment and crisis, but its influence (or the interaction of human will with Pluto's influence) appears to create a signature interplay between these two experiences, which in turn creates a recognizable pattern in our lives of empowerment-crisis-empowerment. By this I mean that Pluto's position often correlates to how an individual or generation is "empowered" in a particular way, and how by the very wielding of this power, crises are created which threaten the goals or even the survival of this same individual or generation. Depending on the choices those in crisis make, this crisis will either overthrow those who brought it upon themselves or those in crisis will emerge from it more powerful and more mature than before. Choosing wisely in a Plutonic crisis seems to consistently involve integrity, while poor choices seem to be defined by integrity's absence. The chapter on Pluto's Generational Influence describes this interplay at some length and detail. I once again beg the reader's patience for my delay in providing real-world examples of this dynamic, as I continue to lay down the broad strokes of my observations.

## The Hero's Journey

It may be a little easier to think of the empowerment-crisis-empowerment dynamic I've described as the Hero's Journey each of us is called to. As a literary trope, the Hero's Journey is easily outlined: The hero is a fortunate individual who loses something he treasured. He must then go on a journey to reclaim it, during which he encounters one or more threats not only his life but the very meaning of his existence. His ability to overcome these threats and return from his journey a different and better person than he was before is what makes him a hero, since "lesser men" (i.e. non-heroes) would not have survived the experience. Modern fictional examples of the Hero's Journey might include Harry Potter, Frodo from the Lord of the Rings trilogy, or Iron Man. More timeworn examples include Aeneas from the Aeneid, Odysseus from the Odyssey, and Gilgamesh. Those who have emerged victorious from their own Hero's Journey can be found in the real world as well as in fiction. Current examples of real people who have famously emerged victorious from their Hero's Journey include Nelson Mandela and Temple Grandin.

One reason Pluto is important astrologically is because it tells us that the Hero's Journey is not just for exceptional individuals and fictional characters. We all have a Hero's Journey we need to make in our lives, and Pluto shows us what that journey will look like: What qualities we have that give us power over our fellows, what we will lose that will send us on our Journey, what we will have to do to get back what we lost, what will be our greatest perils during our quest, and how we can emerge from it all as semi-divine Heroes. Pluto shows us all this on both an individual and collective level.

While Pluto may describe *how* a person—or generation—will ultimately succeed or fail in their Hero's Journey, Pluto does not appear to determine *whether* they will. Even though the Hero's Journey Pluto shows us often seems fated in some way, the results of the process are totally up to us. It's entirely our individual choice whether we emerge as the hero, the villain, or just another casualty in our own and in our generation's Hero's Journey. When properly understood, Pluto can tell us how to make the right choices at critical times in our lives in a way no other Planet can.

As noted earlier, one of the disappointing features of many astrologers' treatment of Pluto is how it is labeled a "generational" Planet, and thereby its relevance in an individual's natal chart and personal life is largely ignored. Observation indicates that Pluto is extremely relevant in describing the lives of individuals, and therefore Pluto's influence on the individual will be the subject of the next chapter.

# Chapter 2: Individual Influence

### Pluto in the Signs and Houses

The first thing Pluto's placement in a natal chart shows is how the native finds empowerment. The form this empowerment takes will be indicated primarily by the Sign; whatever the Sign seeks or embodies—its theme—is what will empower those born with Pluto in that Sign. For example, those born with Pluto in Leo will find that validation (Leo's theme) will be the fountainhead of their empowerment, no matter what the native's Sun Sign happens to be. For example, when one is a member of the Pluto in Leo generation, there truly is power in the validation provided by positive thinking and being constantly told how good one is. Secondarily, where in life the native is most likely to find their font of empowerment will be indicated by the House. So, for example, if Pluto is in the 6$^{th}$ House (the House of service and health) of a Pluto in Leo person, they will likely find that the validation and its attendant empowerment that they seek lies in employment, volunteer work and good health.

Pluto will also indicate how and where crisis is most likely to manifest in an individual's life. This is shown primarily by the House placement, secondarily by the Sign placement, the mirror-image of empowerment. Taking our Pluto-in-Leo-in-the-6$^{th}$-House example, as the native is empowered through validation he may forget that service (part of the 6$^{th}$ House's theme) is what brings this validation to him. Validation is likely to create arrogance (a potential dark side to the validation theme) as it empowers him, causing his initial success in being of service to be eventually crippled by an increasing callousness toward those he is supposed to help. This causes him a loss of ability to be of real service to others and hence to get the validation he needs, though he is more likely to think that any problems that arise lie with others not respecting his great accomplishments rather than with his own arrogance and complacency. His arrogance may eventually cost him his clients and good reputation if it is not checked in time, precipitating a crisis in the very area of life that has empowered him.

In another scenario, this placement could mean that in her search for health (also part of the 6$^{th}$ House's theme) as some kind of a status symbol she can use to lord over others, someone with Pluto in Leo in the 6$^{th}$ House is often deceived or overreaches herself regarding her health, creating the very sickness or injury she was trying to avoid and so inviting the scorn of others rather than their admiration. The fad diet backfires, the exercise is overdone, or her insistence that she knows better than her doctor leaves her blind to a fatal malady until it's too late.

The following are two tables listing the Signs and Houses, along with the themes I use to describe each of them.

| Signs | Theme | | Houses | Theme |
|---|---|---|---|---|
| Aries | Action | | 1st | Self-expression |
| Taurus | Comfort | | 2nd | Resources & Values |
| Gemini | Variety | | 3rd | Daily Routine |
| Cancer | Safety | | 4th | Home |
| Leo | Validation | | 5th | Creativity |
| Virgo | Conformity | | 6th | Service & Health |
| Libra | Fairness | | 7th | Partnership |
| Scorpio | Control | | 8th | Mysteries |
| Sagittarius | Exploration | | 9th | The Wide World |
| Capricorn | Responsibility | | 10th | Life Calling |
| Aquarius | Detachment | | 11th | Hopes & Dreams |
| Pisces | Empathy | | 12th | The Invisible |

These themes are very superficial descriptions of very complex phenomena, and are meant as a summary introduction to the meanings of each Sign and House. That being said, using the cursory information in these tables to match the Sign and House placement of Pluto in the individual's natal chart can provide a basic understanding of how Pluto manifests in one's life.

To sum up the example of Pluto in Leo in the 6th House, the native is empowered (Pluto) through validation (Leo) from being of service and/or being healthy (6th House). Conversely, crisis (Pluto) tends to occur in one's service activities and/or health (6th House) through expecting unearned validation (Leo).

Below I've given more specific descriptions of what I've observed regarding Pluto in the various Signs and Houses. Since I'm limited by a human lifespan, I've only personally observed Pluto in detail in adults of a few Signs at this point—the Signs of Leo, Virgo, and Libra—and cursorily in a couple of others (Cancer and Scorpio). Therefore I'm confining my detailed observations of Pluto in each House to those Signs I've observed in detail—Leo, Virgo and Libra—providing only a more general description of Pluto in Cancer and Scorpio without going into those Pluto Signs' House placements. I will not attempt to describe in detail the impact of Pluto in any other Signs in this book.

**Pluto in Cancer**

The most recent long Pluto generation (over 21 years) in living memory, the quest for safety—Cancer's primary concern—was what empowered this Pluto generation, as well as what formed the basis of their crises. Pluto in Cancer defined the Greatest Generation, the one that bore the brunt of the fighting in the Second World War. Making the world "safe" (Cancer's theme) for themselves, their families and their countrymen brought both empowerment and crisis to members of this generation. The difference between those who were successful and those who weren't is—with 20/20 historical vision—fairly obvious. Those like the fascists and Stalinists who used xenophobia (one of Cancer's most dangerous flaws) as the basis of their power to first conquer and then mercilessly slaughter those unlike them, brought crises upon themselves that they could not survive. In contrast, those such as the United States who used their power based in nurturing others (one of Cancer's more noble traits) to conquer their enemies of the time—Italy, Japan and Germany—and then nurture these former enemies back to health proceeded to overcome their crisis and prosper. Another way of putting this is that those of this generation who cultivated xenophobia and paranoia (expressing the Cancerian need for safety without integrity) to empower themselves at the expense of nurturing those less fortunate (expressing the Cancerian need for safety with integrity) were overwhelmed by their generational crisis, their social movements crushed, and their value-systems largely condemned despite their early successes. In contrast, those of this generation who focused on nurturing those less fortunate, so expressing their Cancerian need for

25

safety with integrity, overcame their generational crisis and handed on a more peaceful world to the rest of us. Despite the negative Cancerian traits of paranoia and xenophobia expressed in such social movements as McCarthyism and the Red Scare in the U.S., this generation in the U.S. became heroes to many who came after them because they drew their power more from the positive traits of their Pluto Sign than the negative.

All powerful governments that sprang from this generation promised a more peaceful world to their followers, but only those who had the integrity to take action to make the world safer for all (even those unlike themselves) such as in the U.S., had the integrity to rise above or overcome their crises.


## Pluto in Leo

For those with Pluto in the gregarious Sign of Leo, both empowerment and crisis revolve around validation. Compliments, applause, the belief that they are vitally important and being the center of others' attention are the things that empower this generation, and those with Pluto in Leo are gifted with both a metaphorical granite pedestal on which to stand and an unwavering spotlight under which to perform. However, being elevated on their pedestal makes it hard for them to see others' point of view, and being constantly in the spotlight blinds this generation to its own silliness and imperfections. Since the spotlight never leaves them, as a group, those with Pluto in Leo come to believe that they can do no wrong, and that what is a priority for those with Pluto in Leo is a priority for everyone. These delusional beliefs bring unjustified self-righteousness, which in turn brings ridicule and censure from others, and the condemnation or indifference that flows from this ridicule and censure is what creates crisis in the lives of those with Pluto in Leo. The bulk of the Baby Boom generation was born with Pluto in Leo. This is the generation who fought for their "right" to peace, justice and equality all over the world in their youth, and who forty years later all over the world fight for their "right" to entitlements at the expense of their own children. In America and Europe, this generation sends its children to fight unjust wars on their behalf in Iraq and Afghanistan. A generation that once broadly fought for peace, justice and equality now largely promotes war, injustice, and inequality, their self-righteousness apparently blinding them to their hypocrisy. Having the integrity to earn the love of others, rather than demanding their "respect", is what will allow those with Pluto in Leo to overcome their crisis.

### 1st House

Someone with Pluto in Leo in the 1st House may seem unrealistically dramatic, utterly predictable and yet somehow mesmerizing when met in person. Their need for attention is likely to exhaust those around them and cause quite a bit of friction over time, but the soap-opera of their life will likely always attract a wide audience. Pluto in the 1st House in any Sign is a likely indicator of chronic and severe health-problems, or of constantly running the risk of injury for the sake of attention. These risks to life and limb can be avoided by avoiding risky situations, but it is unlikely someone with Pluto in the 1st House will be content to do so. Those with Pluto in Leo in the 1st House find that regularly overcoming crisis is what gives them the attention and validation from others that they need, and so the risks they run they run gladly.

### 2nd House

Pluto in Leo in the 2nd House indicates someone who wants to be recognized for their ability to find value. People with this placement have a gift for discovering diamonds in the rough, which they then display as symbols of their own worth. It is likely that those with this Pluto placement will succeed in becoming the talk of the town with their smart investments and attractive discoveries, and this is how the native will find the validation that empowers them. The native's fortunes, however, may prove to be unstable. Greed, envy and dishonesty (either the native's own or that of others) may be hounds always nipping at his or her heels, threatening to take away everything they've worked so hard to gain. Overspending, over-indebtedness, theft and fraud often rob a person with Pluto in Leo in the 2nd House of their well-earned treasures. However, as long as a person with this Pluto placement avoids dishonesty, snobbery and vengeance, there will always be more diamonds for them to find.

### 3rd House

Most of us need the attention of others on a regular basis, but those with Pluto in Leo in the 3rd House may need it like most people need to breathe. They often set up their daily routine to make themselves a spectacle. This is the placement of a freak-show exhibit or door-to-door salesman; someone who needs to be the center of everyone's attention all day every day and so sets up their daily routine to ensure that this happens. Someone with this placement can get the validation they need, but they have to make sure they vary their audience; playing to the same crowd day after day will become stifling for this performer and boring for his or her audience. Having the integrity to be worthy of their audience's attention is what will allow those with this placement to overcome their crises.

## 4th House

Someone with Pluto in Leo in the 4th House needs a home environment that supports him or her totally. There may be an early death or some other crisis involving one or both of the native's parents, and these crises reinforce the native's need for an unconditionally supportive home environment. Those with Pluto in Leo in the 4th House may have a very hard time enduring children in their home, because the child and the native will be in constant competition for attention and validation. People with Pluto in Leo in the 4th House often seek solitude when what they need is intimacy. They are often unreasonably afraid of admitting to others what they truly need. Their fear-based inability to ask others to meet their needs is what causes those with this placement to substitute solitude for intimacy, and this disconnect then precipitates crisis as their need for intimacy fails to be met. Overcoming their fear of vulnerability and sharing their real needs with those closest to them is what will allow those with this Pluto placement to gain integrity and overcome crisis.

## 5th House

Those with Pluto in Leo in the 5th House find validation through their creativity. Whether this creativity is expressed through dramatic performance, a craft or some kind of design-work, or through children, it is creative expression that will bring those with this placement the validation from others that they need. This validation may not translate into monetary reward, however. Those with Pluto in Leo in the 5th House may expect the world to materially support them in their creativity, and their crisis may come through this expectation not being met. Learning how to create for others is what will allow those with Pluto in Leo in the 5th house to overcome their crisis.

## 6th House

Having others applaud their efforts toward greater service and better health is what empowers those born with Pluto in Leo in the 6th

House. They may overcome great hardship in their health or in coming into their service, and expect others to recognize these feats as the grand accomplishments they are. However, those with Pluto in Leo in the 6$^{th}$ House may come to believe their successes in these areas justify self-righteousness later or in other areas of their life, and this self-righteousness is what will precipitate crisis for those with Pluto in this placement. Being of service to others on those same others' terms is how those with Pluto in Leo in the 6$^{th}$ House can overcome crisis.

### 7$^{th}$ House

Pluto in Leo in the 7$^{th}$ House requires that all partnerships—romantic ones, business ones, even enemies—validate the importance of the native. Someone with this Pluto placement may firmly believe that you are who you know, and set out to know only the best and the brightest that they can find. It's always helpful to find oneself in good company, but those with Pluto in Leo in the 7$^{th}$ House may struggle with the fact that they have to earn a place among the beautiful people if they want to remain with them. Accomplishing things on their own is what will allow those with Pluto in Leo in the 7$^{th}$ House to overcome crisis.

### 8$^{th}$ House

In any Sign, Pluto in the 8$^{th}$ House indicates empowerment through exploring the unknown. This exploration could be as a scientist, a member of the police, a metaphysical practitioner, or a researcher of any kind. It can also mean that exploiting the intimacy of others (such as working in the sex trade) is a path that will become available to the native. Revealing things that are hidden, explaining that which is difficult to understand, or using sex to dominate is how those with this Pluto placement are likely to find the validation they require. Revealing too much to others, however, will probably precipitate crisis for those with Pluto in Leo in the 8$^{th}$ House. In believing others want to know all that they have to reveal, those with this Pluto placement may reveal more than their audience is ready for, and so draw condemnation and threat rather than the praise they were seeking. Not everyone is ready for all knowledge, and those with Pluto in Leo in the 8$^{th}$ House would do well to bear that in mind.

### 9$^{th}$ House

Those foreign and exotic things that intimidate many people are what empower those with Pluto in Leo in the 9$^{th}$ House. Exploring the ins and outs of travel, higher education, and religion, then explaining the joys of these things to others is what will recharge a person's batteries when they have this Pluto placement. Their desire to take "the road less traveled" may cause them problems, however. Crisis may come in the form of tragedy while traveling such as disease, accident or kidnapping, joining a cult, or being taken in by a popular academic or religious craze which proceeds to consume much of the native's life. Or, others who take the native's advice may experience these things. Embracing the fact that there is risk as well as reward in exploration (and having the integrity to accept the risks oneself rather than foisting them onto others) is what will likely avoid or mitigate crisis for those with Pluto in Leo in the 9th House.

### 10th House

Being responsible and being loved for it is what empowers those with Pluto in Leo in the 10th House. People with this Pluto placement will likely shoulder more than their share of responsibility, or take on more duties than others are willing to in their quest for respectability. They may expect a lot more respect for doing this than others are willing to give them, however, and the disappointment this can cause may be at the root of any crises they encounter. Knowing when to say "no" to others, and doing things for their own merits rather than for the respect they are supposed to provide, will probably extricate those with Pluto in Leo in the 10th House from whatever crisis they find themselves in.

### 11th House

Pluto in Leo in the 11th House indicates a strong attachment to the future, and to being recognized for helping bring a particular future about. Getting recognition from professional peers or from large swathes of the public for one's accomplishments will likely be very important to those with this Pluto placement. This desire for professional praise may cause crises in the native's personal life, however. They may tend to ignore those close to them for the sake of furthering their public success, or they may choose their friends and lovers unwisely by allowing others in too quickly after two brief an acquaintance. Making time for friends and family (as well as for one's long-term goals) is what will resolve the crises people with Pluto in Leo in the 11th House experience.

### 12th House

While those with a Leonine Pluto in the 4th House may believe they need solitude, those with Pluto in Leo in the 12th truly do. Solitude and self-denial will likely be a source of empowerment for those with Pluto in Leo in the 12th House, but this solitude must somehow bring the native validation and admiration from others. This is a placement for those of the Pluto in Leo generation who discover early in their lives that they need to make personal sacrifices for the greater good. However, this self-sacrifice may be undermined by an expectation of admiration or acknowledgement, and when those things don't materialize as expected, those with Pluto in Leo in the 12th House may feel at a loss. Learning how to let go of these expectations and simply enjoy his or her own company is what will allow those with this Pluto placement to overcome crisis.

**Pluto in Virgo**

This is a very interesting Pluto generation. About half the Pluto in Virgo generation is considered to belong to the Baby Boom generation, and about half are considered to belong to Generation X. Conformity is what empowers those born with Pluto in Virgo, and so this lack of an immediately apparent generational identity actually serves this generation well. Those with this Pluto placement often seek to take on the traits of those with Pluto in Leo or Pluto in Libra to better fit in. This generation profits greatly by keeping its collective head down, its collective mouth shut and its collective nose to the grindstone. However, this very conformity is what precipitates crisis for those with Pluto in Virgo. The self-indulgence of both the Pluto in Leo and Pluto in Libra generations is easy to imitate initially for those with Pluto in Virgo, but the fastidiousness of Virgo eventually rebels at the laziness and excess that comes naturally to both Leo and Libra. Allowing others to make the rules they live by sets the stage for victimization, powerlessness and becoming hyper-critical of these other Pluto generations. Acting on any of these negative feelings then in turn opens up those with Pluto in Virgo to criticism and ostracism, and the inability of those with Pluto in Virgo to endure these things is the essence of their crises. Attention to detail and a need to be of service are two traits that empower those with Pluto in Virgo while setting them apart from other generations. Having the integrity to assert these qualities (even if it means potentially alienating others) is

31

what will allow those born with Pluto in Virgo to overcome their crises.

## 1st House

While all those with Pluto in Virgo may go to great lengths to fit in, those with Pluto in Virgo in the 1st House may be particularly skilled at appearing normal, whether they are or not. This veneer is likely to work well for those with this placement for accomplishing short-term goals, but it is also likely to undermine their intimate relationships, which may keep many of their long-term goals out of reach. Those with this House placement may have to keep in mind that trust can't be bought, it has to be earned. Looking "the part" and talking "the part" may get you into a club or a good job, but it doesn't—and shouldn't—grant real entry into the hearts of others; only proving that you are a trustworthy person does that. When those with Pluto in Virgo in the 1st House are willing to earn the trust of those around them (rather than trying to buy it with words and appearances that are familiar to the other), their relationships blossom and their crises fade away.

## 2nd House

Keeping up with the Jones' may be particularly important for someone with this Pluto placement. This position connotes an obsession with form, which can lead someone with it to always be impeccably dressed and accoutered. It can also indicate someone who is extremely critical of others who do not go to the efforts they do to fit in. This can cause much resentment, and since it is possible someone with this placement may often have to shift their source of income (and so shift the group of people they want to fit in with), this resentment can cause a lot of problems. Remembering that fashion is a game and that possessions are there to serve us—rather than us being there to serve possessions—will help those with Pluto in Virgo in the 2nd House to keep an even keel.

## 3rd House

Attention to detail is part of the daily routine for those with Pluto in Virgo in the 3rd House. Their ability to see and analyze the fine details of themselves and their environment may seem boundless. If this focus on detail is used for learning and for the service of others, the world will be made a better place. However, if this talent is used for selfish reasons or just to feel superior to others, someone with this

Pluto placement is likely to experience a paralyzing paranoia, as they constantly second-guess themselves and others. Also, Pluto in Virgo in the 3$^{rd}$ House indicates that short-distance travel is likely to be a source of drama, whether of epiphanies or tragedies. Either way, it's important that people with this placement pay attention when they commute around town, and drive carefully.

## 4$^{th}$ House

Someone with this Pluto placement may feel that one or both of their parents have written a script for them by which they are bound to live. The fear of making mistakes is probably at the root of this feeling of being bound. A person with Pluto in Virgo in the 4$^{th}$ House will likely feel very attached to at least one if not both of their parents, although this attachment may cause quite a bit of drama or upset in the person's life. Although parents do deserve their child's love, the kind and level of attachment those with this Pluto placement have with their parents is ultimately unhealthy. Children must discover for themselves what they need and how to live, however painful the separation is or however many mistakes are made. Even if the parents really do choose well for their children, a child is lost in the world once the parents are dead and gone unless they have learned how to grow and survive on their own.

## 5$^{th}$ House

Creativity may become a source of never-ending drama for those with Pluto in Virgo in the 5$^{th}$ House. When it comes to any creative activity—artistic pursuits, romance, children—the devil is in the details for those with this Pluto placement. This placement confers great talent in one or more forms of creative activity, but there may be such an intense focus on perfection in the creative endeavor that opportunities are lost or relationships are destroyed. It may be important for those with Pluto in Virgo in the 5$^{th}$ House to learn to not only live with but to love the imperfections in their artistic endeavors, their dating lives, and their children.

## 6$^{th}$ House

This Pluto placement confers an almost superhuman ability to concentrate on a task, and is also likely to indicate someone with gifts in the areas of health and medicine. These abilities often lead to tunnel-vision and workaholism, however, and these traits can then seriously handicap a person's health and/or ability to be of service. Moderation in all things is what will help forestall or lessen crises in the lives of people with this Pluto placement, but they may have a hard time finding it on their own. Someone with Pluto in Virgo in the 6th House may require someone (or a group of someones) in their life to tell them when enough is enough, that working 12 hours a day every day is not healthy and remind them that life is a spectrum of different experiences, not a straight line leading from failure to success.

### 7th House

People with Pluto in Virgo in the 7th House are always looking for the perfect partner, whether in business or in romance. People with this Pluto placement have a great talent for determining what those around them want and need, and for finding ways to provide it. However, people with Pluto in Virgo in the 7th House may set themselves up for public embarrassment through their partnerships, especially their romantic ones (although any business partnerships will likely be quietly successful). For those with this Pluto placement, it may work out better to cultivate an open relationship with someone less than perfect, than a secret relationship with someone unavailable who otherwise embodies perfection.

### 8th House

Ambitious, analytical and self-reliant, those with Pluto in Virgo in the 8th House are driven to succeed in whatever they do. They may obsess about their goals and feel that the ends justify any means in order to attain them, putting a strain on their reputations and relationships, and putting their desired success at risk when (not if) these behaviors and beliefs come to light. Resisting the urge to overturn particularly large and slimy stones will probably benefit those with this Pluto position considerably. Although people with this Pluto position are likely to be very dominant sexually, they are also likely to pride themselves on being attentive to their partner's needs.

### 9th House

This placement tempts those with it to crusade and try to change the world, though the influence of Virgo makes this a much quieter and well-ordered affair than when Pluto is in other Signs. Foreign people and other cultures affect those with Pluto in Virgo in the 9th House deeply. These experiences are likely to inspire those with this Pluto placement to try to bring perfection to the world (whether that means eliminating or disseminating these new ideas) in a highly organized and practical way. This may rebound on those with

Pluto in Virgo in the 9th House as not everyone whose life they try to change will have the same needs or desires as those with this Pluto placement. Walking one's talk while taking a "live and let live" attitude is what will probably benefit those with Pluto in Virgo in the 9th House the most.

## 10th House

Someone with Pluto in Virgo in the 10th House will almost always seem to have their ducks in a row. This is someone who can organize not only themselves but any person or organization they encounter, and this impressive bent for organization will likely open doors and win recognition for them. However, the insistence on organization that comes naturally to those with Pluto in Virgo in the 10th House may make surprise events much harder to deal with than they are for others. Leaving room in one's plans for the unanticipated is the way for people with this Pluto placement to deal with crisis.

## 11th House

Friends, peer groups and professional organizations are likely to provide the greatest empowerment to people with Pluto in Virgo in the 11th House. Reforming, refining or otherwise perfecting the lives of these friends or the organizations of these groups is what is likely to inspire the deepest feelings and greatest efforts from those with this Pluto placement. It is likely people with this Pluto placement will have a powerful and positive impact as a result of their efforts, but at some point they may be forced to move on. Their friends may leave or die early, or professional organizations may accept the help but repay loyalty with abandonment, and as a result those with Pluto in Virgo in the 11th House may find themselves set adrift without compass or companionship. Taking some time out from their reformist zeal to enjoy who people are and where they're currently at will stand those with this Pluto placement in good stead.

## 12th House

Those with this Pluto placement ask Big Questions, and won't rest until they've found equally Big Answers, preferably with footnotes and references. Solitude serves them well for their question-and-answer activities, but they may be inclined to avoid social situations or hide their emotions when they feel threatened. They may also indulge in some kind of hidden sexual relationship or overactive fantasy life over the course of their lives. This avoidance of potential entanglements may seem reasonable to those with Pluto in Virgo in the 12<sup>th</sup> House, but is likely to provoke others to slam doors in the native's face without the native realizing what is happening, or why. Limiting but not eliminating their alone-time is the way out of crisis for those with Pluto in Virgo in the 12<sup>th</sup> House.

## Pluto in Libra

The Pluto in Libra generation wants others to buy their drinks and pay their rent, and is amazingly good at convincing others to do just that. Those with Pluto in Libra are empowered by fairness, in the sense of both "beauty" and "justice". While they are blessed with an abundance of attractive qualities—leisure-time, nice toys, peace to enjoy these things in, physical and social grace—which are often unearned gifts, an obsession with perpetuating their boutique lifestyles often distracts this generation from facing ugly truths and making painful decisions. A tendency to purchase "goods" (physical objects, services, or the aesthetically beautiful) at the expense of "rights" (the integrity of self or others) is what creates crisis for those born with Pluto in Libra. This is Generation X, cynical about the world but willing to put up with its imperfections as long as there's someone to subsidize their life-style. They are aware of all the world's problems but dither as to what to do about it. Their indecision is based in an unwillingness to put their life-style at risk to fight for what's right. Decisiveness and the willingness to sacrifice for a good cause are what will resolve crises in the lives of those born with Pluto in Libra.

### 1<sup>st</sup> House

Those with Pluto in Libra in the 1<sup>st</sup> House are extremely charismatic. The way they are able to attract the positive attention of others can seem almost magical, and those with this Pluto placement rarely have difficulty getting others to do what they want if they can meet or speak directly with the person. This personal magnetism is also what precipitates crisis, however. It can become so easy to influence people into doing what the native wants that he or she never really learns how to take care of his or her own needs or establish their own resources independent of those closest to them. Any time that they must go it alone, natives with this placement tend to become lost and unable to cope. Learning how to be of service to others without having to look good doing it is what will allow those with Pluto in Libra in the 1<sup>st</sup> House to overcome crisis.

### 2<sup>nd</sup> House

Buying beauty comes very easily to those with Pluto in Libra in the 2<sup>nd</sup> House. People with this Pluto placement find it natural to own and be surrounded by the nicest things and the prettiest people, and may feel uncomfortable in anything less than the best clothing, car or neighborhood. The niceties those with this Pluto placement consider necessities others consider luxuries, however, and this difference of opinion is what will often precipitate crisis for those with Pluto in Libra in the 2<sup>nd</sup> House. Overspending and disruptions to income are very common for people with this Pluto placement, and since accumulating savings is often a low priority, a lack of income can overturn the native's apple-cart very easily. Cultivating relationships over possessions, and saving for the inevitable rainy day, is what will allow those with Pluto in Libra in the 2<sup>nd</sup> House to deal with their crises most effectively.

### 3<sup>rd</sup> House

Travel and self-education is what is likely to create turning-points in the lives of those with Pluto in Libra in the 3<sup>rd</sup> House. This will often be a good thing, as the person with this Pluto position will discover more of who they are, make new helpful acquaintances and develop new resources whenever they move or research a topic on their own. Not all of these turning points are going to be positive, however. People with Pluto in Libra in the 3<sup>rd</sup> House are prone to accidents when commuting or otherwise traveling around town, and sometimes what they learn is how good they had it before their last move or that sometimes ignorance really is bliss. Making sure they don't burn their bridges is what will allow those with this Pluto placement to get through their times of crisis.

### 4<sup>th</sup> House

People with this Pluto placement often benefit greatly from their relationship with one or both parents. Those with Pluto in Libra in the 4<sup>th</sup> House are family folk, finding much benefit in following their parents lead or advice. They are often anxious to make a home and family of their own, as a way of living out the role their parent(s) have given them. Over-dependence on their parents and unresolved childhood issues are what is likely to precipitate crisis, however. No one—not even a parent—can fully anticipate the needs of someone else, and eventually the person with Pluto in Libra in the 4th House has to make their own way without the help or guidance of their parents. Also, those with this Pluto placement may believe they must dominate their home and/or blind themselves to anything negative about their parents, because of some form of parental abandonment early in their lives. Facing a situation where life fails to follow the script they got from their parents, failing to make or keep their home beautiful, or being faced with unpleasant facts about their parents may all precipitate crisis for those with Pluto in Libra in the 4th House. Forming and maintaining close relationships with those they are not related to by blood or marriage, and remembering to be fair to themselves as well as others will be very helpful for those with this Pluto placement when their crises arrive.

### 5th House

Beauty, romance, art and children are what empower someone with Pluto in Libra in the 5th House. A person with this Pluto placement is likely to be gifted in all these areas, and will pursue these interests at the expense of other more mundane considerations. This purposeful ignorance of the more mundane aspects of life is what will likely precipitate crisis. Overspending, overconfidence and ignoring responsibilities for the sake of having a good time eventually places those with Pluto in Libra in the 5th House in a position where they can no longer pursue those things that they most value. Finding a way to be of service to others while following their Muse is what will show the way past any crises someone with this Pluto placement is likely to encounter.

### 6th House

Regularly being of service to others, in a way that involves making things more beautiful or people more healthy (a beautician, interior designer, architect, aroma-therapist or reiki practitioner), is what is likely to empower someone with Pluto in Libra in the 6th House. Although people with this Pluto placement may have undeniable talents in these areas, overwork and a tendency to proselytize their belief of "beauty *uber alles*" may sap their physical energy and undermine the good that they do. Finding time to relax and learning to live and let die are what will likely help those with

Pluto in Libra in the 6th House weather whatever storms life sends their way.

## 7th House

Partnerships are where people with Pluto in Libra in the 7th House tend to accomplish the most in their lives. People with this House placement are extremely dominant while remaining sensitive to their partners' needs in any partnership they form, and business partnerships in particular are likely to be successful for them. However, the very success of their partnerships is likely to be the vehicle by which crisis arrives. A partnership involving someone with Pluto in Libra in the 7th House may become more successful or expand its scope much more quickly than either the native or the partner is ready for, forcing one or both of them to rearrange their lives or make some difficult choices based on the partnership. Learning how to make partnerships a true cooperative effort is what will avoid or mitigate crisis in the lives of those born with Pluto in Libra in the 7th House.

## 8th House

Self-reliance and exploring the unknown is what is likely to empower those with Pluto in Libra in the 8th House. This independence and the obsessive interests the native is likely to entertain may put some serious strain on intimate relationships, however. Dominance issues in intimate relationships (especially sexual ones), and the native's need to explore the unknown may be both a source of worldly success and contention with any sexual or business partners the native has. It is often unwise for those with this Pluto placement to share property with someone they partner with, whether a business partner or sexual partner. The ability to let go is what will often serve those with Pluto in Libra in the 8th House best when crisis strikes.

## 9th House

Joining a group or cause that tries to make the world a more beautiful or a more just place will likely bring some form of personal empowerment to those with Pluto in this position. Crusading for these causes may seem natural, but should be avoided. It's very likely that doing so will cause someone with Pluto in Libra in the 9th House to judge people or situations that they don't really understand, and this in turn will yank the rug out from under those with this Pluto position when they act on their judgments. Remaining a spectator to the courtroom drama that is life—rather than trying to take on the part of the Judge—will probably allow those with Pluto in Libra in the 9th House to avoid most crises, and deal with those that arise with aplomb.

### 10th House

People with Pluto in Libra in the 10th House have tremendous leadership ability. They often know exactly what to say to groups of people to inspire and motivate them. The pedestal others put those with this Pluto position on after exercising their singular talent may prove to be precarious, however. Those with Pluto in Libra in the 10th House may find their personal flaws elicit a lot more censure than those same flaws in others do, since groups of people hold those with this Pluto placement in higher esteem and put them in places of greater responsibility. Humility and holding oneself to the same standards as others is what will get those with Pluto in Libra in the 10th House through their crises.

### 11th House

Loyalty to others is what will most likely empower those with Pluto in Libra in the 11th House. Those with this Pluto placement will make a lasting impression on their friends and have a positive impact on the world around them through their loyalty and dedication to whatever cause inspires them. They are particularly susceptible to abandonment and betrayal, however, especially from those friends and causes that they've served so faithfully. Cultivating their blood-ties and hobbies other than their chosen cause will stand people with Pluto in Libra in the 11th House in good stead when their crises arrive.

### 12th House

Solitary contemplation of the Big Questions of life is what will likely bring success and empowerment to those with Pluto in Libra in the 12th House. Those with this Pluto placement are blessed with an ability to discover truth by employing their imaginations, but living too long or too deeply in the realm of the mind may cause them to lose touch with others. This inability to connect with others may cause those with Pluto in Libra in the 12th House to conceal feelings from others that would be better expressed openly, especially feelings of anger or sexual attraction, and these repressed feelings may then precipitate crisis in the form of an inability to create or maintain stable relationships of any kind. Balancing their alone-time with intimate interaction with others is what will allow people with Pluto in Libra in the 12th House to avoid or withstand their crises.


## Pluto in Scorpio

Empowerment "with a vengeance" is likely to describe the Pluto in Scorpio generation very well. Empowerment may come late in life or against great odds to this generation, but when it does come it is likely to make the inequality between the Pluto-Leo and Pluto-Virgo/Libra generations seem insignificant. This is the Millennial Generation (or Generation Y), the children of Generation X supplemented by a second round of child-bearing from the Baby Boomers. Although they far outnumber both the Libra and Virgo generations (approaching the numbers of the Boomer generation) the high unemployment rates that have been hitting the Pluto in Scorpio generation especially hard all over the world for several years have effectively disempowered them early in their adult lives. This disempowerment may serve as the basis for long-term problems and resentments for this generation. Scorpio is the Sign of vengeance and the natural home of Pluto, and Pluto may give this generation the means to strike back against the unjust actions of its elders unless those elders can quickly make themselves part of the this generation's solution. Enacting unrestrained vengeance on the Cancer, Leo, Virgo and Libra generations (or any other vengeance-worthy group) may backfire on those with Pluto in Scorpio, however. Vengeance may cripple them in important ways that even the great victories they are likely to obtain later will not compensate them for. Having the integrity to practice openness and restraint are what will likely get this generation through its crisis.

I believe that the uprising in Tunisia in December of 2010 through January of 2011, and the other popular uprisings it has inspired (such as in Egypt), may have been this generation's "coming out." I believe these events are likely to prove good illustrations of the Pluto in Scorpio generation's modus operandi. A majority of those who participated in the 2010-2011 Tunisian Revolution were unemployed adults under 25 years old (see "Hijacking the Tunisian Revolution" http://english.aljazeera.net/programmes/insidestory/2011/01/2011121

41

165938708665.html), which means that they were born with Pluto in Scorpio. From an online Washington Post article called "Tunisia's Jasmine Revolution" dated January 15th 2011:

*"Not once in my 43 years have I thought that I'd see an Arab leader toppled by his people. It is nothing short of poetic justice that it was neither Islamists nor invasion-in-the-name-of-democracy that sent the waters rushing onto [Tunisia's President] Ben Ali's ship but, rather, the youth of his country."* (See http://www.washingtonpost.com/wp-dyn/content/article/2011/01/14/AR2011011405084.html)

Poetic justice is a Scorpio specialty, as is letting anger build up for years and years and then striking enemies with great effect when these enemies—and others—least expect it. Ben Ali (the former political head of Tunisia) had been in power since the early 80's, and had been the only head of government those with Pluto in Scorpio in that country had ever known. The same was true in Egypt, of Hosni Mubarak. This generation's resentment toward these leaders had been building literally their whole lives. Whether or not the Tunisian Revolution (and other uprisings) succeeds in empowering the Pluto in Scorpio generation, I think we can expect them to continue to dish out generous helpings of poetic justice all over the world. How events in Tunisia in particular play out may show us all more of what empowers this generation, as well as how their use of that power is likely to create crisis.

### Pluto in Aspect to other Planets

Pluto's influence can be analyzed even more closely for specific individuals by looking at the Aspects Pluto forms to other Planets in the individual's natal chart. An analysis of Pluto's natal Aspects to other natal Planets can indicate what areas of the person's life will be most affected by their empowerment and crisis, what coping mechanisms a person is likely to employ when in crisis (and how effective they will probably be), what personality traits or circumstances are likely to come to the fore during a crisis, and how the native may need to prioritize during crisis in order to best overcome it.

The following are two more lists. The first is of the Planets astrologers generally look at in the natal chart excluding Pluto, along with a theme I use summarizing what that Planet usually represents. The second list is one of commonly-used Aspects, with a theme describing what the Aspect means for an interaction between the two Planets.

| Planet | Theme |
|---|---|
| The Sun | Conscious Manifestation |
| The Moon | The Unconscious |
| Mercury | The Executive Assistant |
| Venus | The Comfort Zone |
| Mars | Overcoming Obstacles |
| Jupiter | The Free Lunch |
| Saturn | The Reality Check |
| Uranus | The Awakener |
| Neptune | Intuition |

| Aspect | Theme |
|---|---|
| Conjunction | Reinforcement |
| Sextile | Opportunity |
| Square | Compromise |
| Trine | Gift |
| Quincunx | Missed Expectations |
| Opposition | Feedback Loop |

To continue with our example from earlier in this Chapter, let us say someone has Pluto in Leo in the 6th House. This means they are empowered by validation in their service or health, their service or health also being the locus of their crises. Let us also say this person has Jupiter square to Pluto and Mercury trine to Pluto. This tells us that the things that come most easily to them will compromise their ability to validate themselves through their work or good health, and so relying on those things that come easily as a coping mechanism will probably not serve them. Their empowerment through validation via work or health is represented by Pluto and its placement, what comes easily to them is represented by Jupiter, and the fact that the two are related by the need to compromise between them is represented by the square Aspect. It also tells us that their organizational skills are gifts when helping them attain the validation they need at work and in health, and in resolving crises in these areas. Organizational skills are Mercury's bailiwick, the validation through work and health is once again represented by Pluto and its placement, and the fact that the one is a gift to the other is represented by the trine Aspect.

Below I've listed what I've specifically observed about other Planets Aspecting Pluto in the natal chart. "Positive" (easy or helpful) Aspects include conjunctions, sextiles, and trines, while squares, quincunxes and oppositions are usually considered to be "negative" (difficult or obstructing) Aspects.

When giving the general description of the meaning of each Planet's Aspects to Pluto, I've included the significance of the absence of Pluto's Aspect to that particular Planet as well as its presence. When looking at Pluto in the natal chart, it can be just as informative to see which Planets Pluto does not Aspect as which Planets it *does* Aspect.

**Pluto-Sun Aspects**: This set of Pluto Aspects indicates a personal relationship with power. A person with positive Pluto-Sun Aspects will usually see themselves as innately powerful individuals in some way, and this perception will usually be shared by those the native comes into contact with. This perception will often be based on the native having overcome difficult circumstances or made significant accomplishments early in life, which have given them skills or other qualities that allow them to succeed over the course of their lives. Those with positive Pluto-Sun Aspects will usually be empowered by crisis, even if crisis is something they are uncomfortable with. They may often seem to embody Neitzsche's dictum: "Whatever doesn't kill us makes us stronger." Negative Pluto-Sun Aspects often indicate a person who sees themselves as a victim of the power of others, who has to always ride another's coattails in order to succeed. They view power as something outside of themselves rather than as something they can possess or aspire to, and their coping mechanisms often involve trading their integrity for the sake of short-term gain. Whether the Aspects are positive or negative, those with Pluto-Sun Aspects often feel that access to the truth is their fundamental motivation, and when the truth is hidden from them, may feel that they are failing at life, no matter how outwardly successful they may be.

People with Pluto-Sun Aspects are often instinctively or even compulsively secretive, while they tend to work diligently and successfully at ferreting out the secrets of those people or subjects that hold their interest. Possibly because of these traits, Pluto-Sun Aspects do not appear to be helpful in bringing the native any sort of worldly fame or recognition, since the glare of publicity often comes at the price of one's own private life (where those with Pluto-Sun Aspects often see their real power residing). When the private life of someone with a Pluto-Sun Aspect is curtailed, this may create a fear in the native that they will be cut off from "the truth" that they so treasure. While those with a Pluto-Sun Aspect may certainly be capable of achieving fame, it may be an overwhelming desire for privacy that prevents many of them from doing so.

Natives with a Pluto-Sun Aspect may have a hard time seeing others as equals, preferring to see them as teachers, students, obstacles or aids. This attitude that every relationship must be either dominant or submissive can be a major impediment to healthy relationships, and those with Pluto-Sun Aspects are well-served by learning to relate to others as equals.

Those without major Sun-Pluto Aspects don't see power as something to worry about in their day-to-day lives, and their interactions with power (and their responses to crisis) tend to have a whatever-works quality to them. Secrecy and privacy are also things that those without Pluto-Sun Aspects feel they can take or leave as the situation demands, and they are not likely to view every relationship as necessarily either dominant or submissive. The flexibility possible when Pluto-Sun Aspects are absent is often a great aid in finding fame or other recognition.

**Conjunction**: A native with this Aspect may not be able to imagine a world that does not bow to their wishes, while at the same time see the story of their lives as a tragedy. They are the kings and queens of their life-kingdoms, upon whom their crowns always seem to rest heavily. Whether or not they see themselves as empowered, they make choices and take action as though failure were not even a possibility, seeming to take no notice of how difficult their chosen tasks may be or how small others think the chances of success. This almost-unconscious self-confidence makes many things possible for those born with a Sun-Pluto conjunction that is not possible for others, but this self-confidence can go disastrously wrong when these natives do finally bite off more than they can chew. This self-confidence can also create behavior that is very off-putting to others, if the Sun-Pluto person shoves others and their needs out of the way in the pursuit of their goals. Insensitivity—not only to the needs of others but also to their own—for the sake of what they want is what will often lead to or perpetuate crisis for those with a Sun-Pluto conjunction. Cultivating patience and empathy is what will strengthen those with this Aspect, allowing them to avoid, mitigate, or undo crisis in their lives.

**Sextile**: Those with this Aspect have a positive gift for always being able to see how current problems are also opportunities. People with Sun-Pluto sextiles may see difficulty as a necessary ingredient of success, and find it natural to treat obstacles as ladders to be climbed rather than impediments to progress. This ability to transmute negative experiences into positive ones may make it difficult for them to understand or empathize with those who lack their perspective or their talent for situational alchemy. Empathy may not come easily to those with a Pluto-Sun sextile, and this lack of empathy may eventually present problems that the native is unable to turn to his or her advantage. Patience with others and working toward others' goals along with their own is what will allow those with a Pluto-Sun sextile to deal best with their crises.

**Square**: Those born with Pluto square their Sun often feel that power is something outside of themselves. It may often seem that others hold all the cards in any game the native chooses to play, and that victory always requires some kind of deal with the devil. While it is true that compromise is vital in order to accomplish things for those born with this Pluto Aspect, they often believe that they cannot trust others to give them support or assistance, and this mistrust leads them to misunderstand the nature of the compromises they need to make. This misunderstanding leads to crises in the nature of frustration and degradation for those with a Pluto-Sun square. The less those with this Aspect are willing to trust, the more frequent and the more powerful their crises will tend to be. Trust in themselves and in others is what will lead to the strength these people need to avoid, mitigate, or undo the crises that arise in their lives.

**Trine**: People with Pluto trine their Sun often find the greatest gifts in their lives disguised as tragedies. Very little keeps people with this Aspect down, and almost everything in their lives makes them stronger. A Pluto-Sun trine gives a person a great sense of self and sense of purpose, which the native may find very tempting to use as a bludgeon on others and on difficult situations. Attacking others with these gifts may indeed yield great short-term results, but will leave the native stunted emotionally and sacrifice his or her long-term goals for the sake of short-term gain. Narrow focus is a danger for people with this Aspect, and so a broadening of focus—being able to take in others' feelings and goals and work for them—is what will enliven and strengthen those with a Pluto-Sun trine. So strengthened, these natives will not just be empowered by crisis, but learn to transcend both the crisis and the power that flows from it.

**Quincunx**: Power will not work as expected for those with a Pluto-Sun quincunx. The process of attaining power will likely be quite different than what the native thought it was going in, and the native's wielding of power to attain a particular goal is unlikely to attain that goal. Those with this Aspect gain strength by leaving the seeking and wielding of power to others, and concentrating on expanding their own consciousness.

**Opposition**: People born with Pluto opposite their Sun tend to have a love-hate relationship with power. They often wield power very effectively for short-term gain when they have it, yet they tend to lose any power they possess very quickly. Usually, this loss occurs because the native wields power very selfishly. The native will often be blind to his or her selfishness, however, and project his or her own selfishness onto others. This projection then creates a series of power struggles over the course of the native's life as he or she tries to take back what they see as theirs. This ceaseless conflict creates crisis. The fact that they are almost always on the morally wrong side of power-struggles prevents the native from attaining their long-term goals and from building stable relationships with others. Cooperation and the ability to acknowledge their own selfishness when it arises is what will build strength and integrity, and so end crisis, for those with Pluto opposite their Sun.

**Pluto-Moon Aspects**: Any Aspect from Pluto to the Moon in the natal chart is both very powerful and very disruptive. It indicates a high probability of negative and compulsive patterns of behavior that are very difficult for the native to break. The native is likely to have urgent needs that can only be satisfied by the exercise of power, whatever this exercise may cost themselves or other people. Compulsions the native doesn't fully understand that involve exploiting others often dominate the lives of those with natal Pluto-Moon Aspects. People with these Pluto Aspects are often blessed with a charisma and animal magnetism that serves their immediate appetites very well, but makes developing integrity (and therefore overcoming their long-term crises) equally difficult. They often project an unassailable outward confidence to the world, which hides an overwhelming fear of inadequacy.

Those with Pluto-Moon Aspects may have a very intense and very difficult life-long relationship with their mother, while having a very limited or non-existent (though less fraught) relationship with their father. Making peace with their mother, or at the very least acknowledging their grievances to themselves and airing them with their mother, will probably be necessary for those with Pluto-Moon Aspects to find the integrity to overcome their personal Plutonic crises.

People with Pluto-Moon Aspects often have to fight harder than most to develop self-awareness. Also, crisis is something those with Pluto-Moon Aspects often seem to embody rather than overcome.

As a side note, every autistic person's natal chart that I've seen has a Pluto-Moon Aspect. More research is needed on this topic. Whether or not Pluto-Moon Aspects prove to be an indicator of autism, all Pluto-Moon Aspects do appear to indicate a tendency toward obsessive behavior.

Although often problematic, the single-minded passion and the ability to embody crisis those with Pluto-Moon natal Aspects often display can be directed into literally world-changing accomplishments. The lives of Gandhi and George Washington are two well-known examples of what Pluto-Moon energy is capable of when channeled in a positive way. Although both men embodied the crisis of the time and place that they inhabited, they became symbols of what could be accomplished in the midst of crisis. By their heroic examples they led others out of their collective crisis and—literally—into new worlds of freedom and independence.

Although they have their benefits, Pluto-Moon Aspects do not appear to be an aid to becoming famous. They often represent critical obstacles the native has to overcome within him or herself on their way to success, rather than strengths that the native can always play to. They also make it difficult for the native to connect emotionally with others, often because the native's own emotions are so overwhelming. Worldly success and empathy are both possible for those with Pluto Moon Aspects, but in attempting to attain these things they must often overcome obstacles others do not have to contend with and which others may have a hard time understanding.

Those without Pluto-Moon Aspects often have an easier time developing self-awareness and looking at feelings objectively.

**Conjunction**: For those with Pluto conjunct their Moon, power may somehow be wrapped up in or embodied by their mother. While the native's mother will likely always be there for him or her, at least as long as the mother is alive, those with a natal Pluto-Moon conjunction may become very limited physically and emotionally through their reliance on their mother. These limitations will likely be the root of other seemingly unrelated crises that they experience. Overcoming long-term crisis in the lives of those with this Aspect will often require creating and maintaining their independence from their mother, or in taking care of their mother later in life. Finding the strength to do this will also allow them to become fully adult themselves, and to become good parents in their own right.

Crisis may be a constant part of the life of a person with a Pluto-Moon conjunction. When reacting to a stressful situation, those with a Pluto-Moon conjunction may be prone to exhibit extreme or compulsive behavior, because their natural coping mechanisms will often involve expressing their subconscious emotional nature rather than engaging their rational intellect or awareness of external circumstances. This disproportionate and irrational behavior will almost always get them what they want immediately at the expense of their long-term stability and credibility. Learning to respond to situations rationally (rather than reacting to them emotionally) will probably best serve those with a Pluto-Moon conjunction long-term.

**Sextile**: With this Aspect, the mother often provides needed opportunities for the native to reach his or her goals. Reliance on the mother is probably more helpful than harmful for those with a Pluto-Moon sextile. Although this need for the mother may persist throughout the native's life, the more independence they can gain from their mother the better. There are things that their mother will never be able to help them with, and the earlier the native realizes this and starts developing their own resources, the better equipped they will be to deal with crisis when it materializes.

Those who have a Pluto-Moon sextile tend to immediately respond to a stressful situation with withdrawal and a wait-and-see approach to discover the best resolution. This tactic will probably serve the native well, and is something that should be cultivated.

**Square**: People born with a Pluto-Moon square may often have to fight the influence of their mother in order to overcome their crises. Whether their relationship with their mother is a pleasant or unpleasant one, what they learned or got from their mother is probably not going to help those with a Pluto-Moon square when their back is against a wall. In fact, their mother's actions or influence may be the very reason why their back is against the wall in the first place. People with this Aspect are likely to realize this fact early on and start developing other coping mechanisms to deal with crisis, but there may be some lingering and often unconscious resentment toward the mother for this effective abandonment. Acknowledging and overcoming this resentment is what is probably needed by those with a Pluto-Moon square to find the strength to master any difficult long-term situations that plague them.

Stress or immediate crisis may be paralyzing to those with this Aspect. This paralysis is likely to stem from a profound mistrust of their own reactions and coping skills, and from a fear of making mistakes. It may be important for those with this Aspect to remember that no one is born with great coping skills, and that most of us have to learn by making mistakes. Acting decisively when the pressure is on is what will help those with a Pluto-Moon square deal with both short-term and long-term crisis. When under duress it may help those with Pluto-Moon squares to remember that making mistakes is a necessary step toward success, not a guarantee of failure.

**Trine**: When Pluto trines a person's Moon, their mother is likely to be one of the best things the native has going for him or her. People with this Aspect often (but not always) have a lot of love for their mother, and will trust their mother's advice and example above all others. Even those with a Pluto-Moon trine that have a rocky relationship with their mother will benefit from the relationship, even though it may mean simply learning the kinds of people and behavior to avoid through their mother's example. As with all other Pluto-Moon Aspects, however, it is important for the native with a Pluto-Moon trine to leave their mother's orbit. Those with a Pluto-Moon trine may be dangerously complacent about relying on their mother, and be blindsided when their mother's beneficial influence is gone or otherwise fails them. Developing their coping skills beyond reliance on their mother, while their reliance on their mother still works, is what will help those people with this Aspect avoid or mitigate crisis.

Stress may be something people with a Pluto-Moon trine find that they have an emotional need for. Their reactions to stressful situations are likely to be quite good, and they may feel at loose ends if the level of stress in their lives falls too low.

**Quincunx**: A Pluto-Moon quincunx indicates that the native and his or her mother may often seem to be ships that pass each other in the night. They may both have a hard time understanding each other, and it's likely neither of them will meet the other's expectations no matter how hard they try. The native may waste precious time or other resources in trying to understand or accommodate the mother, and that is likely to be the root of the long-term crises the native experiences. When the native realizes that he or she is not going to be able to satisfy mother (or the expectations that he or she may have subconsciously absorbed from mother) and gets on with the business of living his or her own life, seemingly unrelated long-term crises in the life of the Pluto-Moon quincunx native will appear to resolve themselves.

Those with a Pluto-Moon quincunx may often be taken by surprise by stressful events, and their emotional reactions are likely to be inappropriate to the situation, such as feeling happiness when someone dies or feeling depressed about succeeding. However, if they learn to respond consciously rather than react emotionally to an immediate crisis, those with this Pluto placement will often cope quite well.

**Opposition**: This Aspect may indicate a real love-hate relationship between the native and their mother. The native is likely to strongly desire independence and to make their own way, but circumstances may seem to conspire against the native maintaining that independence. The likely reason for these setbacks is a strong unconscious desire the native has for validation or support from the

51

mother, which wars with the conscious desire for freedom from mother's expectations or other restraints imposed by mother. Since this struggle for independence is likely to inform much of how the native lives his or her life, many of the native's long-term crises are likely to involve this inability to fully separate from mother. To overcome long-term crisis for those with a Pluto-Moon opposition, what is often required is an acknowledgement of their strong desire for their mother's love. With this acknowledgement, a more honest relationship with their mother is possible. If the needed love and support cannot be found from their mother, the native with this Aspect will know that they need to consciously seek the same from other relationships. In this way, the native will still get the support they need to maintain their independence or accomplish and maintain their other goals.

Like those with a Pluto-Moon conjunction, crisis may be a constant in the lives of those with a Pluto-Moon opposition. The coping mechanisms of those with a Pluto-Moon opposition may often mitigate immediate crisis by ignoring or prolonging the problem, because those with this Aspect are reluctant to look at the part they themselves play in creating crisis in their lives. Asking themselves what they want from a person or situation long-term (rather than just right now) when things become stressful is what will probably best serve those with a Pluto-Moon opposition.

**Pluto-Mercury Aspects**: These natal Aspects indicate that others' perceptions of the native and the use of physical abilities are primary mechanisms for the native's empowerment. While this often translates into success as an entertainer, sports figure, or sometimes politician, it is not often seen in the charts of those successful in other fields. Pluto-Mercury Aspects seem to denote an ability to inspire and entertain others, and sometimes to persuade them rather than directly command their obedience.

A common theme with Pluto-Mercury Aspects is that while others' perceptions are a source of empowerment to these natives, this empowerment is accompanied by a difficulty for the native in understanding intellectual points of view other than the native's own. Intellectual objectivity is a trait those with any Pluto-Mercury Aspects would do well to cultivate, since narrow-mindedness of one sort or another is often at the root of their long-term crises.

Those without Pluto-Mercury Aspects may have a slightly easier time keeping an open mind.

**Conjunction**: This Aspect indicates a person that can make him or herself believe whatever they need to, to get by. A person with a Pluto-Mercury conjunction may often have a very idiosyncratic way of speaking, but yet be very verbally persuasive nonetheless. When crisis arises, these people may be more skilled than anyone else at sticking their heads in the sand and convincing themselves—and others—that ignoring the problem will make it go away. This is clearly a very dysfunctional coping mechanism, but it may be curiously difficult for those with a Pluto-Mercury conjunction (or those under the spell of the native) to recognize this. Objectivity and doing things for themselves—rather than convincing others to act for them—is the best way to cultivate integrity for those with a Pluto-Mercury conjunction, and is therefore the best way for them to overcome the long-term crises in their lives that their rationalizations, obliviousness, and reliance on patsies tends to create.

**Sextile**: Their mind and their ability to communicate is what will open doors for those born with Pluto sextile Mercury. "News they can use" is the tool those with this Aspect will most often find available to them to accomplish their goals and overcome immediate obstacles. Their reliance on their own intellect and verbal communication abilities (and those of others) may not help them recognize deeper emotional currents in themselves and their surroundings, however, and this deficit is what will probably precipitate crisis for those with this Aspect. Learning how to empathize and respond to others' emotional needs is what will allow these natives to overcome their crises.

**Square**: The words and ideas of others may often be intimidating and disempowering for those born with Pluto square Mercury. Natives with this Aspect may have some form of language barrier or deficiency in their education that gives them a feeling of inadequacy when dealing with those who they see as being more articulate or highly educated than they. Embracing diversity in thought and speech is what those with Pluto square Mercury need if these inadequacies are ever to be resolved. The native must recognize that both they and others are entitled to their own thoughts and their own ways of communicating, that there is no one "right way" to think or speak. It will probably be up to the native to build bridges in these areas when bridges need to be built, rather than rely on others to do so. This bridge-building may require quite a bit of humility from those with this Aspect, but if they are willing to take the chance they will almost certainly find it is well worth the effort.

**Trine**: A Pluto-Mercury trine indicates a person very comfortable with communication and intellectual activities of all kinds. Natives with this Aspect thrive in any kind of intellectual or communication-based situation, and will place themselves in these situations as a way to deal with most of life's challenges. As with other Pluto-Mercury Aspects, however, those with this trine may have a very hard time truly understanding any point of view other than their own, despite their quick minds and excellent communication skills. They may see thoughts and communication as being one way—from themselves to others—and have a very hard time seeing the value of two-way communication. This blind-spot may create crises in their intellectual models, communications and relationships with others that their usual coping mechanisms can't address. Travel may be one of the best ways for those with this Aspect to broaden their perspective, giving them the intellectual openness needed to overcome their long-term crises.

**Quincunx**: Those with this Aspect may find that words never quite communicate what the native thinks they do. This means that the effect their speech has on others is usually different from what they originally intended, and that they often misunderstand the ideas of others. Sometimes this disconnect works to their advantage and sometimes to their detriment, but either way the native may need to keep an open mind and a very loose set of expectations when dealing in the realms of ideas or words.

**Opposition**: The words and ideas of others may be empowering in the short-term and very limiting in the long-term for those that have Pluto opposite their natal Mercury. Repeating the words or promoting the ideas of others may bring much short-term success to those with this Aspect, but the native may come to feel that their own words or ideas are overshadowed by this "expert opinion" from outside of themselves. This feeling of disempowerment may eventually lead the native to lash out or otherwise engage in self-destructive behavior, perhaps without the native being fully conscious of why he or she is really doing so. Spending time on developing their own ideas and insights is what will empower those with a Pluto-

Mercury opposition over the long-term, allowing them to promote both themselves and others in the best possible way.

**Pluto-Venus Aspects**: These Aspects speak of how comfortable the person is with crisis. This doesn't necessarily indicate whether they deal with it well or poorly, just whether or not being in crisis is a comfortable place for them. Some people with positive Pluto-Venus Aspects are so comfortable being in crisis that they never learn how to get out of it, while some with negative Pluto-Venus Aspects are so uncomfortable with crisis that they quickly learn how to resolve or avoid them. Pluto-Venus Aspects also often indicate that beauty is a very important part of how they experience crisis, whether as a way to deal with it, create it, or both.

Pluto-Venus Aspects will also describe what role women tend to play in crisis for the native, and whether spending money is likely to help or hinder the native when dealing with crisis. Positive Aspects indicate that beauty, women and spending money will be helpful, and negative ones indicate that beauty, women and spending money will be a hindrance. Pluto-Venus Aspects do not appear to play a significant part in acquiring fame.

Those without Pluto-Venus Aspects will find that beauty, women and spending money do not play any consistent role in creating, averting or resolving the crises they face. Pluto-Venus Aspects appear to be strangely rare in the charts of royal personages.

**Conjunction**: Luxury is likely to be both a major source of empowerment and the root of most crises in the lives of those with this Aspect. Those with Pluto conjunct their Venus are often very good at creating and embodying symbols of "the good life" for themselves and those around them, and using these symbols to get others to do what they want or to make themselves wealthy. Crisis will often manifest when the native overspends or when others discover just how much maintenance is required to retain the glamour the native with this Aspect provides. Although problematic, a native with this Aspect is likely to accept these crises as the cost of their lifestyle, and so be disinclined to end crises that arise. They are often afraid that an end to their crises would mean an end to the life of luxury that they prize so highly. Whether the native is male or female, the women in the lives of those with this Aspect will tend to be very beautiful or otherwise glamorous while the native's relationships with them will be very expensive to maintain. Women are likely to be a source of both empowerment and crisis for those with a Pluto-Venus conjunction. To overcome long-term crisis in their lives, those with a Pluto-Venus conjunction must find the strength to trade form for function. Instead of insisting that they be or have the most glamorous or the most beautiful of them all, those with this Aspect must learn to separate their wants from their needs, and make sure the needs of themselves and others are met before catering to their wants.

**Sextile**: Those with a Pluto-Venus sextile are likely to be very talented at getting other people to spend money. It's also very likely that this talent will be used for good, in that both the native and those the native convinces to spend money will profit greatly through the money spent. Also, women will tend to provide opportunities for the native to find new and better ways for the native to spend (other people's) money. Crisis will most often occur for these natives as a side-effect of their success. Those with a Pluto-Venus sextile may come to embody success, which may become twisted in some people's minds as a caricature of this same success. People with this Aspect are often blessed with a very healthy approach to crisis. They often see crises as the cost of doing business and so are ready and willing to deal with them, and have a knack for minimizing or shortening crisis. A Pluto-Venus sextile will often give the native a gift for turning what others may see as a problem into an opportunity. Those with this Aspect often hold the opinion that any press is good press as long their name is spelled right, and that the slings and arrows fortune hurls at them today are launched by the customers and donors of tomorrow. As long as the native works toward enriching others along with him or herself, crisis is something they will naturally deal with well.

**Square**: Spending money is often anathema to those with this Aspect, and while there may be a great desire for female company, the women in their lives are often more of a burden than a help. Romance in particular is a very painful experience for those with a Pluto-Venus square, as they find that their romantic expectations are rarely met. All of these apparent storm-clouds will indeed rain on the native's parade, but they do have silver linings the native can make use of. By learning to spend occasionally on quality rather than often in quantity, the native with a Pluto-Venus square will probably discover that they have a better understanding of long-term value than their peers. This understanding gives them a true gift for profitable long-term investment. Relations with women (including romantic ones) can also be made positive through moderation; as long as the native recognizes that they often have unreasonable expectations regarding romantic partners and/or women in general, and makes sure his or her needs are met in other ways before partaking in the desired interaction with women, relations with women will remain balanced and enjoyable. People with this Aspect will often find their greatest wealth, success, and happiness later in life. Those with a natal Pluto-Venus square are extremely uncomfortable with any form of crisis, and so they may spend more time than most understanding crisis in their own lives as a way to avoid it as quickly and as thoroughly as they can. This may sometimes make them seem overly pragmatic or pessimistic to others, but their problem-solving skills are often phenomenal, especially in their later years.

**Trine**: Surprisingly, this Aspect often presents more problems than solutions, as the desires of the native indicates by Venus are bent towards Plutonic phenomena. This Aspect indicates an unhealthy fixation on possessions, usually of an expensive or dangerous nature. This fixation may be so powerful that the native will sacrifice their real needs, their decency or even others on the altar of their desire. It also indicates that women will tend to have a very powerful and unhealthy influence on the native. This fixation on possessions and the negative influence of women are the two things most likely to be at the root of crisis in the native's life. Developing positive relationships with men and learning to value positive relationships over expensive possessions is what will prevent or mitigate crisis for those with a Pluto-Venus trine.

**Quincunx**: Those with a Pluto-Venus quincunx often find beauty and even profit or recognition in what other people consider to be odd or ugly. Money will often not satisfy the needs of those with this Aspect, and they must accept that others will not usually share their aesthetic tastes. Relations with women will often be difficult for those with this Aspect whether the native is male or female, due to seemingly insurmountable miscommunications and misunderstandings. Accepting their own tastes as being normal for themselves, and taking a live-and-let-live attitude toward others' tastes—even in the face of intolerance—is what will give those with this Pluto Aspect the strength to overcome their crises.

**Opposition**: While natives with this Aspect are often capable of creating great beauty for others or discovering the beauty in others, they may seem incapable of creating it for or discovering it within themselves. Their ability to bring beauty and comfort to others is likely to be great, and so others are likely to overlook (at least for awhile) the glaring deficiencies these natives are likely to exhibit. People with Pluto opposite Venus may see beauty and even comfort as something outside of or beyond themselves, and this belief may lead to phobic or pathological behavior that prevents these natives from attaining the comfort or physical beauty their contemporaries take for granted. It is unlikely women will play a large or positive role in the lives of those with this Aspect. Self-acceptance may be very difficult for those born with Pluto opposite Venus, but it is self-acceptance and the realization that they deserve the beauty and comforts others enjoy that will allow these natives to overcome long-term crises in their lives.

**Pluto-Mars Aspects**: Those with Pluto-Mars Aspects often see life as a series of competitions, which they are determined to succeed at. Any Aspects from Pluto to Mars indicate a generally excellent problem-solving ability, yet one with an integral Achilles Heel that sends the native tumbling into failure if the native over-relies on it. The House and Sign Mars is in will show what area of life and what tools the native will tend to utilize to try to address their crises. The Sign and House *Pluto* is in will often identify related traits that aid the native when used in moderation but become their Achilles Heel when relied on too heavily. For example, Ross Perot has a sextile from Mars in Taurus in his 11th House to Pluto in Cancer in the 1st House. He saw many long-term problems in the United States that he felt could not be resolved through "business as usual" (i.e. a candidate from either of the two mainstream U.S. political parties) and tried to resolve these problems at one point through a highly idealistic run for the U.S. Presidency. He felt that using his personal money (Taurus) to fund a campaign to directly (Mars) ask the country to allow him to take forceful individual action (also Mars) to enact his long-term vision (11th House) as a way to increase the good of all (also 11th House) was the best way to bring both his own and society's goals about (11th House). However, with his Pluto in Cancer in the 1st House, while his forcefully-expressed individualism (1st

House) and obvious personal power (Pluto in the 1$^{st}$ House) initially helped him get on the ballot nationwide, his equally obvious desire to increase his personal power (also Pluto in the 1$^{st}$ House) was undermined by his inability to relate to those unlike himself (a negative Cancerian trait) and his insistence that everything be done his way (1$^{st}$ House). For Aspects from Pluto to Mars, it is often more useful to look at the House/Sign dynamic of Mars and Pluto as just described, than to try to distinguish the influence by specific Aspect. Because of this, I have not described the specific Aspects from Pluto to Mars in detail as I have for most of the other Planets.

Pluto-Mars Aspects generally also indicate how interactions with men and the native's sex life are likely to affect their ability to deal with crisis. These Aspects may also indicate whether or not earning a higher income will help resolve the crises that come up for the native. Any Aspect (positive or negative) from Pluto to Mars indicates that all three of these things—men, sex, and earning money—will be helpful to the native in attaining his or her goals. However, if the native makes these people or things ends in themselves rather than means toward other goals, then men, sex and earning money will likely become problems rather than solutions.

Pluto-Mars Aspects are common in those who are famous.

These Aspects are found more frequently in the natal charts of entertainment figures, literary figures, famous scientists, sports figures, and recognized visual artists than chance would indicate, as well as being more common than in the baseline population (see the Appendix for specific information). The problem-solving skills and will to succeed implied by Pluto-Mars Aspects seem to serve people well in many fields of endeavor.

Those with no Pluto-Mars Aspects may find that men, sex, earning money and expressing their anger have little to do with how they cause or resolve their crises.

While I have not found there to be much distinction between individual Pluto-Mars Aspects, I have found some useful distinctions between the sets of typically positive and typically negative Pluto-Mars Aspects. Pluto-Mars Aspects will often show whether or not a person's temper is part of their empowerment and crisis.


**Pluto-Mars Conjunction, Sextile, and Trine**

Those with a Pluto-Mars conjunction, sextile or trine will generally find that they can transmute the anger they feel into pure, directed energy, which then increases their endurance and tenacity when focused on a particular goal. Over-reliance on this trait can cause harm to others (and deepen the native's crisis) by causing the native to become overly driven and aggressive; however, with a positive Pluto-Mars Aspect, the native will generally discover this drawback early in life, and learn to use their gift of fury in moderation.

**Pluto-Mars Square, Quincunx and Opposition**
Those with a Pluto-Mars square, quincunx or opposition may often find that their temper tends to get the better of them, and that it leads them into overly violent statements and actions that they quickly have cause to regret. This seems especially true for those with a Pluto-Mars square. Natives with these difficult Pluto-Mars Aspects may always have a very excitable temper and focus too narrowly, too quickly, when trying to resolve problems. However, if they learn how to resolve their problems without resorting to anger or tunnel vision, there is very little that they cannot accomplish.

**Pluto-Jupiter Aspects**: Aspects from Pluto to Jupiter in the natal chart will often indicate that the native will receive quite a bit of empowerment through and recognition for their generosity toward others, and that others will often be unreasonably generous toward them. However this generosity and the apparent good luck this embodies will also probably lead the native into giving more than they can afford to or trying to "fix" people or situations that they don't really understand, and so generosity also tends to be the root of their crises. People with Pluto-Jupiter Aspects are often quite lucky early in life and when making any kind of a beginning, but it is very easy for them to take their initial good fortune for granted and so sabotage their later efforts through laziness, lack of follow-through or ill-considered generosity. While those with Pluto-Jupiter Aspects usually have a lot to give, it is often very dangerous for them to accept charity from others and should avoid doing so. Those with Jupiter-Pluto Aspects who can remember to mind their own business rather than that of others will likely do very well for themselves.

Jupiter is usually considered an extremely positive influence, and normally when it forms Aspects with other Planets even hard Aspects (such as a square) are easier to handle and even pleasant. This does not seem to be true when Jupiter Aspects Pluto, however. In the case of Jupiter and Pluto a trine between the two Planets (usually considered the most positive possible Aspect) is often just as fraught with problems as a square (often considered the most negative possible Aspect), so it is important for those with "positive" Pluto-Jupiter Aspects to not assume that Jupiter will moderate Pluto's tendency to precipitate crisis.

Those with any kind of Pluto-Jupiter Aspect are more likely than others to use violence or coercion to achieve their long term goals. Often, those with this Aspect firmly believe in the premise "The ends justify the means." This belief—and the actions that stem from it—are often at the root of both short-term success and long-term crisis for those with Pluto-Jupiter Aspects, causing them more problems than generosity does. To avoid or mitigate these crises, natives with Pluto-Jupiter Aspects must learn to limit the means they use to achieve their goals to those a majority of others find acceptable.

Despite the potential problems Pluto-Jupiter Aspects denote, this is the single most common kind of Pluto Aspect in the charts of most kinds of famous people. The exception is business-people, who surprisingly have about the same incidence of Pluto-Jupiter Aspects as the general population.

Those without Pluto-Jupiter Aspects will find it more natural to mind their own business, and to understand that the means used to achieve a goal determine whether or not the goal can be reached. Dumb luck and generosity will not typically precipitate crisis in their lives, but they can't count on dumb luck or the kindness of others to get them out of a scrape either.

**Conjunction**: Those born with Pluto conjunct Jupiter often find that they are on this Earth to empower others. The more quickly they embrace this fact, the more successful they are likely to be long-term. This Aspect indicates a person who will probably attain quite a bit of success early in life, and who will get further ahead faster than their peers. However, if those with a Pluto-Jupiter conjunction do not find a way to give away their success (i.e. help others be at least as successful as they are themselves), only with increasing effort will they be able to keep what they have gained early on, and even with increased effort they may be unable to take the final few steps to complete the journey toward their own ultimate goals. If someone with a Pluto-Jupiter conjunction finds that they are stuck in life, the secret to getting unstuck is to pass their current wealth or position on to others.

**Sextile**: People with a Pluto-Jupiter sextile will find that doors are opened for them and opportunities present themselves through the frequent generosity of others, and this means they can depend on the help of others to advance their own career or otherwise achieve their short-term or medium-term goals. However, those with this Aspect often believe in the total rightness of whatever they are trying to achieve, and so will often feel justified in using means that others find reprehensible (such as violence or blackmail) in pursuit of their goals, when persuasion and leading by example fails. This hypocrisy is what usually creates crisis and failure in the lives of those with this Aspect. A willingness to submit their actions to scrutiny, and to make amends when they wrong others, is what will allow those with this Aspect to avoid or mitigate crisis.

**Square**: Interestingly, the square of Pluto to Jupiter is probably the most favorable of the Pluto-Jupiter Aspects. This Aspect often gives the native a true gift for diplomacy. The generosity that is usually indicated by a Pluto-Jupiter Aspect manifests here as a true ability to see other totally different points of view, and in an ability to find common ground between sworn enemies as a way of showing them the benefits of making peace with each other. The danger with this Aspect is that the native can become so enamored of the new and foreign points of view he or she is exposed to, that he or she forgets why they were negotiating or what they were negotiating for in the first place. This means that they often make great and widely-recognized strides early on in their diplomatic efforts, but they risk disappointing their constituencies eventually if they forget who and what they are supposed to be working for. People with a Pluto-Jupiter Aspect are able to achieve almost anything, as long as they can stay focused on what it was that they initially set out to do. Sometimes, those with a Pluto-Jupiter square achieve what they originally set out to accomplish and regret having done so, because their exposure to the opposite point of view has convinced them that their early efforts were misguided. Those with this natal Aspect can avoid this kind of crisis by avoiding making absolutist statements or taking absolutist positions, however alluring as these statements or positions may initially be.

**Trine**: The trine is usually one of the most beneficent of Aspects, so it is ironic that the trine from Pluto to Jupiter is probably the most dangerous of the Pluto-Jupiter Aspects. This Aspect, more than any other Pluto-Jupiter Aspect, will incline the native to violence as a way to resolve problems or achieve goals. This Aspect often gives the native a real talent for reaching short-term goals through violence, but this tactic will put not only their long-term goals but also often their very lives at risk. There are two things that make a difference between success and failure, long life and early death for those with this Aspect: firstly, they must fight not for themselves or for those close to them, but for the wider interests of society. Secondly, they must fight against the minorities who would threaten the interests of the majority, rather than align with them for the sake of advantage. This Aspect can indicate someone who is a crusader and a "true believer," who will stop at nothing and sacrifice everything (including that which rightfully belongs to others) in pursuit of their goals. If they pursue selfish interests or align themselves with "big money" or other "big" interests, they will fail (and possibly die early as part of their failure). If they pursue the wider interests of their society and do so without falling prey to the lure of big money and high-society connections, they will likely succeed and live long productive lives, leaving a good legacy for their children.

**Quincunx**: Those with a Pluto-Jupiter quincunx are often very generous individuals, and may have a talent for putting together charitable organizations or otherwise giving on a grand scale. However, the generosity of those with this Aspect is often most effective on a small scale, rather than a large one. It will often be second nature to those with this Aspect to give to others, and this should be a good thing. To ensure that the impact of their generosity is a net positive one, the native needs to make whatever giving they do immediate and short-term. For those with a Pluto-Jupiter Quincunx, helping a neighbor find a job, or giving them a ride to a new job when they can't yet afford a car, is more likely to have a positive impact than founding charities for large numbers of people the native does not know. If the native insists on giving on the macro level rather than keeping to the micro, his or her personal generosity may be quickly overshadowed by the mistakes or corruption of the supposedly charitable organizations he or she helps start.

**Opposition**: Those with a Pluto-Jupiter opposition are likely to be highly charismatic and theatrically talented individuals, who also have the rock-solid belief that others are there to support them and so can be made to do or believe whatever the native wishes. Although charisma and talent may get these natives pretty far, their attitude of natural superiority and assumption that the rules others must live by don't apply to them will eventually cause them some serious setbacks. Playing by the rules everyone else plays by is what will allow those with a Pluto-Jupiter opposition to not only avoid or overcome crisis, but to learn what love and relationships are really all about.

**Pluto-Saturn Aspects**: Pluto-Saturn Aspects speak of how the native uses the authority of others to empower themselves. Whether the Aspects are positive or negative, working with those in authority or with power structures is what will both empower the native and directly cause crisis in their lives.

These Aspects also show the possibility of empowerment through interactions with the father. Although the father may be very obstructive to the native early in life and a burden on the native's time or money in later years, the native may find it impossible to be at peace with themselves without reconciling with their father before the father has passed on. If this reconciliation does not occur, the native will have a much harder time finding the integrity and strength of character to overcome their crises.

Both crisis and empowerment are things that those with Pluto-Saturn Aspects have to work long and hard in order to realize, and once attained, may be permanent features of the native's life. Objectivity regarding one's goals—asking oneself if a goal desired really is a good idea for the native and for society—is something anyone with a Pluto-Saturn Aspect would benefit from cultivating.

In areas such as business, entertainment, science and sports, these Aspects are not a burden, although they do not appear to be much of a help, either in gaining or keeping success in these endeavors. Generally, those with Pluto-Saturn Aspects welcome the obligations and expectations of society, and so have little wish to change them or abandon them. This makes them solid citizens, but does not usually help the native to become memorable. The presence of these Aspects appears to militate against a career in literature, politics or the arts, probably because the native accepts the status quo.

Those without Pluto-Saturn Aspects usually find that their fathers and authority figures do not play a lasting role in their empowerment or their crises, and their personal process of empowerment and crisis is both more rapid and easier to reverse than for those with Pluto-Saturn Aspects. Pluto-Saturn Aspects are often more useful by their absence than by their presence when it comes to acquiring fame, especially in the fields of literature, politics and the arts.

**Conjunction**: Those with a Pluto-Saturn conjunction often influence others by being cogs in a great machine. Belief in authority is almost a religion for those with this Aspect; their acceptance of government and other forms and pronouncements of authority is so automatic it is often unconscious, and they have a hard time understanding how anyone could not accept authority as unquestioningly as they do. This belief often stems from a very great love for, or close identification, with the father. People with a Pluto-Saturn conjunction do very well as any kind of bureaucrat or other representative of father-figure-like authority, since they accept this kind of authority as an integral part of themselves. They are at a disadvantage, however, when the authorities they believe in are wrong or are being purposely deceitful, since the very idea that "their" authority is capable of error or deceit is totally alien to them. This blindness will often cause them to follow leaders or support policies that work against their own interests and undermine the very authority that the native believes in so strongly. Learning to think both independently and critically, and learning that defying authority for the sake of the greater good is not a betrayal of their father, is crucial for those with a Pluto-Saturn conjunction.

**Sextile**: A Pluto-Saturn sextile indicates that the native's life is often defined through obligation, usually to the family and specifically to the mother. There will often be a single, seemingly inescapable obligation (or a series of such obligations to the mother and/or family) that the native is required to discharge throughout their life. Although this will often feel very restrictive to the native and appear to prevent them from or at least delay their growth as an individual, fulfilling this obligation is truly their higher calling. In

65

fulfilling this obligation, they will usually be blessed with long life (when others close to them die early), good health (when others around them are sickly), wealth, a great legacy and the blessing of history. They can even have the seemingly forbidden relationships or do the supposedly forbidden things they so crave, as long as they are open about their desire for these things and are willing to accept the conditions placed on said things by their overriding obligation. Refusal to fulfill this obligation will not usually rob them of their material accomplishments or of leaving a positive legacy, but it will make their lives much harder. They will often lack financial resources and find it difficult to gain the respect or recognition of others, until and unless they discharge their familial duties.

**Square**: This Aspect indicates a very authoritarian mindset, as a reaction to a very negative experience or set of experiences involving the native's father. The native will often identify strongly with particular forms of authority or authority structures, but gravely misunderstand the nature, capabilities and goals of the authority he or she identifies with. The native may have quite a bit of self-respect invested in being seen by others as an authority figure, but it may be very difficult for them to secure the respect from others they feel they deserve. Unfortunately, others are probably right in not according the native the authority he or she feels she deserves, since those with this Aspect often misunderstand or have a very incomplete knowledge of what they claim to be their area of expertise, and so the greater authority they are given the greater are the mistakes they will tend to make. Those with a Pluto-Saturn square often experience crisis through being ridiculed or brushed aside by others who the native feels need to be impressed, because the native's conclusions were faulty or their predictions were the opposite of the truth. Cultivating relationships with others as equals, rather than seeking ways to master or otherwise impose upon others, is what will address any identity crisis or crisis of rejection those with this Aspect may encounter.

**Trine**: Those with a Pluto-Saturn trine will often find that they benefit from the authority structures built by others. They will often come into their authority through inheritance, when an established leader surrenders the reins (either through death or retirement) and names the person with this Aspect as their successor. Those with a Pluto-Saturn trine have a talent for making an organization grow once they take it over, but they may also have a fixation on death or destruction which will seep into the group and eventually lead it to a very bad end. If there is a negative fixation of this nature, it often is rooted in the lack of a good father-figure in the native's early life. In order to succeed long-term and nurture an organization that will be both positive and long-lasting, the native must learn to embody the positive father-figure that he or she never had, rather than embodying the abusive ones that he or she may have actually had to put up with. If they had a positive experience with their father early in life, a native with this placement will likely benefit from the authority structures of others throughout their lives, but are less likely to inherit a large organization or strive for recognition within one.

**Quincunx**: A quincunx usually indicates a "disconnect" between the two Planets' energies in the native's life, but a Pluto-Saturn quincunx does not fit this mold. In fact, my observations indicate that Pluto and Saturn are at their most harmonious when placed in this Aspect. Natives born with a Pluto-Saturn quincunx have a strong attunement to the passage of time, and he or she will be empowered by cultivating this attunement, as well as by studying time and improving its measurement. It also appears to give the native a very easy interaction with authority in all its manifestations. This is a very rare placement, and while I have observed how it appears to empower a native born with it, I have not yet discovered the method by which it induces crisis.

**Opposition**: People with a Pluto-Saturn opposition often set themselves up as authority figures by portraying themselves as an opposing force to other more established authority figures. This can often give them a small but vocal following, but since their authority is based on negativity it will often be short-lived and destructive. In order to maintain the authority they crave, and to leave a positive legacy for those who come after them, those with a Pluto-Saturn opposition have to give themselves and their followers something positive to live and to fight for, rather than sacrificing everything in their fight against something. Those with a Pluto-Saturn opposition often have very strained or non-existent relationships with their natural father, and reconciliation of any kind may be particularly difficult.

**Pluto-Uranus Aspects**: Pluto-Uranus Aspects show that conformity is a force for empowerment (and non-conformity a source of crisis) in the native's life. People with a Pluto-Uranus Aspect will often find that keeping their head down and their nose clean will be simplest and most direct route to empowerment. They may have a particularly hard time maintaining this seemingly simple course of action, however, because they constantly see the shortcomings of mainstream society and feel compelled to try and leave it. The further they walk on the wild side, the more disempowered and deeper into crisis those with Pluto-Uranus Aspects will tend to fall. Finding the strength to stay within the society and bring new ways to improve it—such as Louis Braille did with his system of writing for the blind, or Michael Faraday did when he upended the physics model of his day by demonstrating the unification of electricity and magnetism—is what will allow those with a Pluto-Uranus Aspect (like the two famous men just mentioned) to avoid or mitigate crisis.

It is very common for those who become infamous to have a Pluto-Uranus Aspect. In trying to find ways to live outside of society's norms they more often find sickness and dysfunction than truly livable alternatives. The grisly parodies of normalcy created by Jim Jones and Charles Manson are extreme examples of what can happen when those with a Pluto-Uranus Aspect look too hard for alternatives to mainstream life.

Because both Pluto and Uranus move so slowly, when they form Aspects to each other there is a 4 to 10 year period when everyone born will have more or less the same Aspect in their chart. Since such a large number of people are born under each Pluto-Uranus Aspect, I find it often more informative to look at Pluto-Uranus Aspects' generational effect rather than trying to view them on an individual basis or by particular Aspect.

When looking at Pluto-Uranus Aspects generationally, the first thing to note is that there was no Pluto generation born in the 20th or early 21st century where every single member was born with a Pluto-Uranus Aspect. However, there is one generation born in the 20th century—those with Pluto in Libra—where this Aspect is completely absent. What this means is that those born in a generation other than Pluto in Libra will have a mix of those who need to conform and those who can safely explore the boundaries of custom and expectation (if they are inclined to do so). The astrology indicates that the Pluto in Libra generation as a generation, however, does not have any members that would hold it back from doing unconventional or extreme things. This doesn't mean that everyone born with Pluto in Libra should leave their families, start a commune and expect to found the next Findhorn. It does mean that those with Pluto in Libra are generally much less limited by conventionality than those born under other Pluto generations. Whether they choose to take advantage of this fact or not is up to them, of course, but the absence of Pluto-Uranus Aspects in the Pluto in Libra generation's charts means that if they individually or collectively feel inclined to duck under the ropes society sets up to keep people in line, those with Pluto in Libra are more likely to return with wonders and entertaining stories rather than with monsters and mournful tragedies.

**Pluto-Neptune Aspects**: Pluto-Neptune Aspects seem to define the "common man" in the 20th and early 21st centuries. When I put together a control-group of 20 randomly selected charts of people born between 1926 and 2008, so I could get a baseline for how often different Pluto Aspects appear, 18 of the 20 charts (90% of the sample) were born with a Pluto-Neptune Aspect. Groups of 10 natal charts of those deemed famous or otherwise successful by society— the wealthy, the powerful, artists, sports figures, even the infamous— born during the same time period show significantly lower percentages of Pluto-Neptune Aspects (from 20% for literary figures to 75% for royalty), except for sports figures, who appear to have about the same incidence of Pluto-Neptune Aspects as the baseline population (90%), and science figures where I could not get enough names to form a usable group. Because these sample groups are smaller than the standard I set of twenty members, and because I was unable to find usable names at all for the Science group, I have not included these sample groups in the Statistical Data Appendix. The other sample groups I did include show the same general lack of Pluto-Neptune Aspects relative to the baseline population, although the specific numbers were slightly different. The point is that while there certainly are wealthy, powerful and otherwise successful people that have Pluto-Neptune Aspects, it would seem that Pluto-Neptune Aspects most often indicate a tendency to remain part of the herd rather than a tendency to stand out from it.

There has been only one Pluto-Neptune Aspect seen since Pluto's discovery—a sextile—that has been in effect on and off from October 1944 until September of 1995. Based on my read of history and what I see of our current culture and world-state, this Aspect seems to indicate a willingness to use fantasy in place of reality when reality does not present us with the things we want or need. It's empowering to be able to make do with what we have, but if we let this coping skill lead us into mediocrity and apathy it becomes an avenue for crisis instead of empowerment. Finding the strength to wake from our dreams and reach for something better in the real world is what will let those of us born with this Aspect avoid or mitigate the crises in our lives.

## Final Comments on Pluto's Individual Influence

Each component of the natal chart already mentioned (Signs, Houses, Planets and Aspects) can and should be analyzed more deeply on an individual basis, to add further clarity to Pluto's influence in an individual's natal chart. Analyzing the Sign and House placements of Planets that Aspect Pluto will also yield more detailed results, of course (as will analyzing Pluto's relationship to various other chart components not mentioned above, such as the Ascendant, the Lunar Nodes, and Chiron), but the material above should be enough of a beginning to demonstrate the usefulness of analyzing Pluto in the individual natal chart.

I have not found any placements or Aspects of Pluto that appear to guarantee success or failure in an individual's life. As stated in the General Significance Chapter, Pluto appears to merely detail the process by which a person succeeds or fails, not dictate whether they will succeed or fail. It appears to me that it is integrity—or its absence—which separates the Heroes from the villains and casualties in our various Plutonic Hero's Journeys.

To once again return to our previously-used metaphor for the purpose of illustrating one form that a Plutonic Hero's Journey could take, someone with Pluto in Leo in the 6th House will usually have crises that manifest at work or regarding health. His efforts in these areas validate him and bring him short-term success, but this very success is likely to be the root of long-term problems. If his problem is with his work, where his early successes in being of service nurtured an arrogance that brought him to eventually disregard the needs of his customers, he has two options: One option is that he can write off his disappointed clients as too stupid to understand him or too weak to knock him from his pedestal, and try to drag more validation from them by fraud or extortion. This path leads to ostracism and eventual disempowerment. Or, he can admit his failings, practice some humility and once again match his service-oriented words with service-oriented action. He may not regain his former position, but even if he doesn't, he will then be able to continue as a contributing member of his community, and so find the validation—and the power—he needs. This will also maintain his good name in the community, allowing him to leave something to the community (and his children) that is likely worth having. If the crises manifest with his health, it may be because he tries to gain the admiration of others through his physical prowess or apparent immunity to disease or age. This tempts him to take "short-cuts" (fad diets, plastic surgery) to apparent good health that helps his ego immediately at the expense of his long-term health. Once again, his options are clear: He can pretend he is superhuman at the expense of a painful old age and possibly premature death, or he can accept the fact he has physical limits, humbly acknowledge them and live within them.

In both versions of the Plutonic crisis described above, the first long-term solution involves turning away from integrity, while the second involves cultivating it. Although it seems obvious when put this way that cultivating integrity is better than not, many of us find that this decision is not quite so clear-cut when facing our own Plutonic crises. As difficult as the choice often is, however, its results are unwaveringly predictable: Cultivating integrity creates a happy ending to our own Hero's Tale, while sacrificing integrity on the altar of expediency is always a devil's bargain, prolonging and expanding the crises we face.

The most important step in successfully resolving any Plutonic crisis is to recognize that maintaining integrity is the foundation of success. When preserving integrity is made the priority in dealing with Plutonic crisis, the crisis will be overcome. In contrast, if integrity is sacrificed for the sake of expediency, to preserve one's other goals or health in the face of Pluto's crisis, ultimate failure appears to be assured. This is probably the most important point in understanding how to resolve a Plutonic crisis, whether on a personal or generational level.

Pluto's is very much a black-and-white, all-or-nothing influence. This makes judging whether a person or group succeeds or fails the "test" Pluto presents a straightforward process. First of all, the evidence of success or failure is not seen in the natal chart, but in the choices the native or a given generation makes. Success scenarios involve a major crisis being faced and defeated. The crisis is defeated when it does not return in any form after being faced successfully, and the native or generation is clearly empowered by this victory, able to do things after winning this victory that were not possible before it was won. Success scenarios also involve leaving some form of favorable legacy, cementing the victory in memory and in history. Failure scenarios involve the attempted avoidance of crisis or the inability to overcome it. When a generation is failing its Plutonic test, its crisis situation becomes permanent or is apparently resolved only to return in a different but recognizable form shortly after each supposed "resolution". Failure is also indicated by the inability to leave a legacy, or to leave one that burdens one's descendants rather than benefiting them.

Once again, although Pluto does not decide our ultimate success or failure on an individual or collective level, it does appear to show the process of how success or failure will ultimately come about. This is another reason why understanding Pluto's influence is so important.

The next chapter will describe how Pluto influences each generation on a collective level, and analyze an example of the successful resolution of a generation's Plutonic crisis.

# Chapter 3: Generational Influence

As the astrologer Dane Rudhyar noted as early as 1951, the Sign Pluto is in when a person is born is perhaps the best possible way of assigning a "generation" to which the person belongs. I agree, as it appears that Pluto's Sign position indicates what advantages a generation starts out with, what they need in order to succeed, and what shared crises they will experience in their efforts to reshape the world to their liking. Looking at Pluto generationally can show how by collective human action the world will be made a heaven or a hell, and explain how one state of being can be transformed into the other. This can give the native a good idea of what group activities and goals would most empower him or her, as well as an idea of which would lead to crisis and so should be avoided.

Pluto often shows how groups of people go about obtaining and wielding power, and how this process creates historical crises that these same groups must overcome or be condemned by history.

### An analysis of Pluto in Cancer's Hero's Journey

Looking at the needs, goals, and actions of the generation born with Pluto in Cancer illustrates Pluto's generational influence very well.

Pluto was in Cancer from roughly 1913 through 1939. Since the theme of the Sign Cancer is "safety," Pluto's placement in that Sign would indicate that safety would be the center of gravity around which would revolve this generation's crisis and empowerment. Born amidst the crises of the First World War, a devastating worldwide plague in 1918 and The Great Depression, this generation clearly developed a Cancerian need for safety. Circumstances conspired to show them that the world was a hostile and unforgiving place, and that banding together with one's family—nuclear, racial or national—was the way to peace and prosperity. Many sought and gained power during this time through promoting the safety and sanctity of race and nation, but always at the expense of "them," those strangers already outside the protective shelter of one's home. The French, English and Poles only felt safe by crippling Germany through the Treaty of Versailles, Russian communists cemented their hold on power through informers and mass-imprisonment of those who disagreed with them, and the Great Depression only ended when nations mobilized to fight each other. This scrambling for the safety of home and family through widespread use of beggar-thy-neighbor (or kill-the-stranger) tactics created worldwide crises that made every group less safe.

And so the major crisis faced by those born with Pluto in Cancer found its culmination in World War II, which began as Pluto culminated (entered the last degree) in Cancer. Members of this generation from all over the world were drafted into various militaries to protect their nurturing shells of culture, country and home as World War broke out for a second time. Millions of people sacrificed, fought and died so that the world would be a safer place for themselves and their descendants.

Only a minority could win this initial violent contest. That minority was what came to be known as the Allied Powers—Russia, the U.S., and the U.K. being the largest—with the U.S. reaping the greatest rewards. The true resolution of the crisis lay not in winning the shooting war, however, but in the choices that were made afterwards to resolve the ongoing crisis of which the War was merely a symptom. Another large-scale shooting war between the U.S.S.R. and the U.S. became quite likely after WWII, after the Soviets went home and licked their wounds for a few years. Military, ideological and technological "races" were all begun as World War II drew to a close, very much as a direct consequence of how the war was ended and how the spoils of war were divided.

However, the most powerful of the victors of World War II—the U.S.—chose to direct its newfound power into exhibiting more of the positive traits of the winning generation's Pluto Sign (Cancer) than its negative ones. They chose to nurture their most powerful enemies during the war—Germany and Japan—into the industrial powerhouses they remain at the beginning of the 21st Century, rather than try to forge a new Treaty of Versailles to keep their former adversaries "safely" out of the way while beginning a new war with a new "them" (the Soviets). The resolution for this generation's crisis could have been better—we did get the Cold War out of it, after all—but by finding the strength to use Pluto to emphasize the positive traits of Cancer of nurturing and mercy, and to de-emphasize Cancer's natural xenophobia and paranoid insecurity, the Pluto in Cancer minority that won the last World War used their power with integrity.

75

By making the world as safe as they could for the largest number of people possible, this generation ensured that another "shooting" world war and thermonuclear war were both avoided.

This is not to say there was no individual or collective xenophobia or paranoid insecurity on the part of the Pluto in Cancer generation in America. There certainly was, as McCarthyism and the Red Scare in the U.S. illustrate. But the time, effort and money—the energy—put into rebuilding America's former enemies (and the compassion and objectivity it took to do that, even if these were alloyed by self-interest) outweighed that which was put into xenophobia and paranoia, allowing the Pluto in Cancer generation to leave the legacy of a more peaceful and prosperous world to its children.

As this example illustrates, resolving a Plutonic crisis appears to involve recognizing that the greatest threat any crisis poses is to our integrity. When cultivating integrity is made the priority in a crisis, the crisis will be overcome and a positive legacy will be left. When integrity is sacrificed to preserve one's self-importance, goals or health in the short-term, ultimate failure is assured. As stated at the end of the last chapter, this is probably the most important point in understanding how to resolve a Plutonic crisis.

Shorter Pluto generations (those of 21 years or less) also appear to have a life-cycle that can be tracked by Pluto's movement through the Zodiac during their lifetimes. I've included what I've observed of this apparent "life-cycle" below. I hope my use of horticultural terms to describe each stage of the Cycle makes it easier for the reader to connect with and understand what I'm describing.

## 7-Stage Pluto Life-Cycle

**Plowing**: *Birth conjunction, while Pluto is still in the same Sign.* Those being born are still dependent on the power of those birthing them at this stage, and are wielding very little power themselves. The important factor is the length of time this stage of the Cycle lasts; the longer it lasts, the more historically significant the generation will be.

**Seeding**: *Semisextile, while Pluto is in the first Sign after the birth-Sign.* These are truly the "formative years" of a given generation, although it may not seem so to the generation itself. Much of the generation's unconscious patterning will form based on reactions to events and experiences that occur during this stage.

**Germination**: *Sextile, while Pluto is 2 Signs from the birth-Sign.* The generation begins to come into its own during this stage. It is a period when doors begin to open for this generation, and they begin to acquire and wield power. The generation begins to learn the rules of how to succeed in life.

**The Storm**: *Square, while Pluto is 3 Signs from the birth-Sign.* This is the Cycle where a generation's decisions are most likely to determine overall success or failure. The rules of success the generation learned during the Germination may seem to change radically during this time, as doors are slammed in their face and some of the power they thought was theirs is taken from them. Strength of character (the ability to maintain integrity in the face of adversity), adaptability and sacrifice are what's required of a Pluto generation during its Storm, rather than power.

**The Harvest**: *Trine, while Pluto is 4 Signs from the birth-Sign.* Whether a given generation ultimately succeeds or fails, this will be its most prosperous time. It may seem that anything the generation touches during this stage of the Cycle turns to gold. It may also seem to members of this generation that the good times will never stop. If the generation has successfully weathered its Storm, the good times will indeed continue to roll into its Withering time. If the generation has failed to weather its Storm, however, the fruits of its Harvest will vanish shortly after the Withering sets in.

**The Withering**: *Inconjunct, while Pluto is 5 Signs from the birth Sign.* This is when the world begins to truly pass a generation by, both culturally and physically. The generation becomes more and more aloof as the world stops making sense to them, and in turn the world takes less and less account of them. Only during the Withering can a generation begin to get an idea of what kind of legacy it is leaving to its heirs, and whether it has succeeded or failed in weathering its Storm. Members of the generation start to die in large numbers during this stage.

**Lying Fallow**: *Opposition, when Pluto is 6 Signs from the birth Sign.* With this Cycle the world has now fully passed this generation by.

The vast majority of this generation will be dead by the end of this stage. Objectivity can be applied to the actions taken and results gained by those of this generation during this stage of the Cycle; this is where the historical perspective on the generation's actions truly begins.

**Example: Pluto in Cancer**

Using the example of the Pluto in Cancer generation to illustrate this 7 Stage Cycle may be instructive.

**Plowing**: *Pluto in Cancer, 1913-1939.* This is when the generation was born. Pluto was in this Sign much longer than it was in the Signs of this generation's children or grand-children, so much of their children's and grand-children's lives were shaped by the choices this generation made.

**Seeding**: *Pluto in Leo, 1937-1958.* Going to war and the rebuilding of devastated nations afterward, seen as a world-wide competition to validate (Leo's theme) particular belief systems or ethnic greatness over all others, and the stability and material prosperity winning this contest brought the victors; these Leonine circumstances formed the unconscious patterning of the Pluto in Cancer generation, guiding them to measure validation for ideals through material success.

**Germination**: *Pluto in Virgo, 1957-1972.* From the late 50's through the late 60's, the Pluto in Cancer generation enthusiastically embraced conformity (Virgo's theme) as the path to success, and they were very taken aback when their Pluto in Leo children didn't seem to feel the same way. As long as Pluto remained in Virgo, however, the Pluto in Cancer generation held the upper hand in both money and politics, since Pluto empowered their conformity to social norms rather than the personal validation that their Pluto in Leo children sought.

**The Storm**: *Pluto in Libra, 1971-1984.* This was a bittersweet time for the Pluto in Cancer generation. Both the bitterness and the sweetness came as just desserts. It was a bitter time for the Pluto in Cancer generation, in that Pluto suddenly began to empower their irresponsible, non-conformist Pluto in Leo children with greater political and economic clout as they strived for social justice for women, minorities and the people of other nations. This meant that for the first time, the Pluto in Cancer generation lost a major conflict (Vietnam), and had to face the fact that success and happiness could not always be measured—or even provided—by material security. It was sweet, in that the Pluto in Cancer generation saw the lasting fruits of its efforts to spread capitalism around the world in the first rumblings of the earthquake that would bring down the juggernaut that was the Communist Threat.

**The Harvest**: *Pluto in Scorpio, 1983-1995.* This was a time when the Pluto in Cancer generation could celebrate its success. Communism and the Cold War died quietly during this stage, thanks in part to the nurturing the successful members of the Pluto in Cancer generation gave to those of its members less materially fortunate around the world. Pluto in Libra, the wild party that scared the conservative Pluto in Cancer generation half to death, morphed into the more socially conservative and materially driven atmosphere of Pluto in Scorpio. This was much more emotionally and materially comfortable for the Pluto in Cancer generation, since this stage was a time when material success once again became paramount and it became clear that there was a distinct and positive legacy that the Pluto in Cancer generation was leaving to its descendants. The Pluto in Cancer generation was justifiably confident during this stage that it could look forward to a comfortable and well-deserved retirement.

**The Withering**: *Pluto in Sagittarius, 1995-2008.* Retirement was in full swing for the Pluto in Cancer generation during this stage. Although they still wielded a powerful voting block in the form of the AARP in the U.S., and prominent members of the generation could speak authoritatively on social and moral issues around the world due to their track record, the Pluto in Cancer generation removed itself further and further from current affairs as Pluto moved through Sagittarius. Their memories of the Great Depression of the 1930's and the wild financial speculation that brought it on were widely ignored in the hysteria of the stock-market, internet and real-estate bubbles, as even members of the Pluto in Cancer generation found themselves swept up in the general enthusiasm. The world was beginning to pass this generation by, but not enough time had passed yet for history to begin effectively judging the impact this generation had.

**Lying Fallow**: *Pluto in Capricorn, 2008-2024.* At this stage, the Pluto in Cancer generation is totally removed from the shaping of world events, and has been for several years. This makes it possible to begin to look back on this generation's contributions with some objectivity, and make definite statements about its traits and its accomplishments. Although not perfect, the Pluto in Cancer generation was largely successful in overcoming its generational crises, and we can only hope that future generations can live up to—if not surpass—their example.

When Pluto leaves Capricorn, its sojourn in various Signs will once again lengthen to between 21 and 32 years. With the time Pluto spending in each Sign taking up a larger portion of people's lives, it remains to be seen if the 7 Stage Cycle will hold.

When looking at various Pluto generations and whether or not they've conquered their generational crisis, it's important to remember that "it's not over until it's over." The other Pluto generations who are alive and in their adult years currently (those with Pluto in Leo, Virgo, Libra, and Scorpio) don't seem to be coping as well as the Pluto in Cancer generation did, but as long as there is life, there is hope. The Pluto in Cancer generation, as successful as they were, could've thrown away their success at any time during their Plutonic Life Cycle, condemning the world to nuclear holocaust or never-ending war between the industrialized nations. By the same token, those of the Pluto in Leo, Pluto in Virgo and Pluto in Libra generations could step forward and claim their success at any time.

**Other Pluto Generational Effects**

Pluto's influence also appears to have other effects on the generational level that are worth mentioning. The longer Pluto is in a Sign, the more likely it is that some kind of widespread crisis—a war or other distinct event such as a plague that kills massive numbers of people—will occur. The world has not experienced a World War since 1939 (Pluto leaving Cancer) or devastating worldwide plague since 1918 (Pluto in Cancer). Given this reasoning, it is unlikely the world will experience another "hot" world war or worldwide plague before Pluto enters the Sign of Aquarius in 2023. Also, it appears that the longer Pluto is in a Sign relative to those around it, the more likely that generation's actions and decisions will have historical significance. The actions of the recent Pluto in Cancer generation are likely to prove more historically significant than the actions of those born with Pluto in Leo, who in turn will probably find themselves more historically significant than those born with Pluto in Virgo,

Libra, Scorpio, Sagittarius or Capricorn. Those generations born with Pluto in Aries, Taurus and Gemini, however, are the ones whose actions are most likely to shape the course of centuries.

There are advantages to being born in a "short" Pluto generation, however. It appears that those born in a Sign Pluto moves through quickly may get short shrift in the history books, but will probably lead more peaceful lives than their more historically significant brethren.

# Chapter 4: Transiting Influence

As illustrated by the Pluto Life Cycle detailed in the last Chapter, Pluto's transiting influence shapes our individual and collective experiences just as powerfully as its natal influence does. This chapter goes into further detail regarding the forms this transiting influence can take.

In a general sense, Pluto's transiting influence shows what "the rules of the game" (i.e. how to gain material success and public recognition) are likely to be at a given time, when they will change, and what they will change into. The theme of the Sign Pluto is in will show what kind of approach will be needed to get ahead at a particular time. The rules change when Pluto leaves one Sign and enters another. Pluto moves too slowly to help stock-market investors time a market, but it does make a good indicator of which markets will do well or poorly over the medium-term. For example, while Pluto was in the lucky and risk-taking Sign of Sagittarius from early 1995 to early 2008 taking risks and trusting to luck—exploration, Sagittarius' theme—was the path to success. Startup capital, stock-markets of all kinds, and derivatives—all traditionally risky forms of investing—came to be considered safe and reliable ways of investing money, since so many people were making money in them so consistently during this time. When Pluto moved into Capricorn later in 2008, however, the rules changed. Although there are still those in 2011 who profess to believe that various markets are still just experiencing a hiccup, startup capital is now very difficult to come by, stock-markets are widely seen as being risky, and derivative financial products are sold at deep discounts or not at all, because no one is willing to buy them any longer. While Pluto is in the pragmatic and generally unlucky Sign of Capricorn, anything that can go wrong will have a much greater likelihood of doing so. Learning how to be responsible and down to earth will be the way to get ahead for most of us while Pluto remains in Capricorn. Not all of Pluto's transitions are quite so memorable, but when Pluto moves from one Sign into another that is in many ways its polar opposite (such as from Cancer to Leo, or from Sagittarius to Capricorn), then its effects are more likely to be dramatic.

### Pluto in the Houses

For specific individuals, Pluto's most noticeable transiting influence can be discerned in the House it is moving through in the individual's natal chart. The area of life represented by the House will probably require constant attention while Pluto is making its transit through it, either because of increased empowerment or constant crisis. Either way, watching for how empowerment creates crisis may help individuals to make the most out of Pluto's motion through their individual charts. Although the rules for success will change for most people at the same time when Pluto changes Signs, the times when Pluto changes Houses (a different time for each individual) will be when the individual's personal life goes through dramatic changes. There may or may not be crisis involved, but the individual's priorities will likely shift quite radically when Pluto moves from one House to another.

The transition of Pluto from one House to another will often manifest as a new and unexpected event or set of circumstances that the native finds highly disruptive. At first, this event or set of circumstances may seem exceptional, and the person may believe (at first) that this is a temporary crisis that their old life-style, beliefs, or relationships will weather. Over the course of a year or so, however, it will become increasingly apparent that this "disruption" is actually a "new normal" that must be adapted to, no matter how improbable or impractical adaptation may seem. Acceptance of the new circumstances is key; despite the initial disruption that usually accompanies it, the new normal isn't usually bad, and often brings with it unforeseen gifts. The more quickly the old life is released, the more quickly the native can get on with the business of living. Trying to hold on to the "old life"—whether that life involved a set of beliefs, a job, a relationship, or whatever other circumstances that served as its foundation—in the face of a Pluto House transit is a futile exercise. When Pluto changes Houses, something in the native's life that seemed permanent (probably brought in while Pluto was in the old House) usually departs forever, to make room for what Pluto represents in the new House. Trying to hold on to or bring back "the dead" is doomed to failure, and attempting to do so prevents us from experiencing the life and regeneration Pluto is bringing us by its movement through the next House.

84

Pluto may move back and forth over a person's House Cusp several times over the course of a twelve month period. The length of time this takes is due to Pluto's frequent retrograde motion, which causes it to enter, leave and then re-enter the new House in the person's natal chart. What this means is that as dramatic as the changes are when Pluto moves from one House to another, it will often take about a year for the transition Pluto is representing to unfold.

I've provided a brief description below of what one can expect when transiting Pluto leaves one House of the natal chart and enters the next.

**1st House**: During the year or so when transiting Pluto enters this House, there is a good chance that hidden undercurrents in one's life are will come to a head, in a way that will change how others perceive the native. If the native has unknowingly alienated someone (or a group of someones) it may come back to haunt him or her in a big way during this transit by costing them a job, a spouse, or another important relationship. This alienation coming home to roost may appear in the lives of those around the native instead of in the native's own, although the broken or altered relationships will almost certainly impact the native's life as well. This might manifest as the native's parents suddenly announcing a divorce or the native's children's previously-unacknowledged problems coming to light very publicly. Greater freedom and healing will likely be made possible through these circumstances, but they may be traumatic when they first occur.

While Pluto is in the 1st House, barriers toward self-expression are likely to fall away. In and of itself this is a very good thing, but there are a couple of potential pitfalls. The main one lies in trying to be too controlling. While this transit is in effect, self-control and self-discovery will prove far more profitable than trying to control others or unearth their hidden motivations. The other potential pitfall involves the probably unavoidable confrontations that come along with this transit, as those people, institutions or ideas that have held back the native's self-expression lose their hold on the native. This confrontation and detachment is a basically healthy process, but the native may need to remember that detaching with love is just as important as the detachment itself.

**2nd House**: While Pluto is in the 2nd House, the native is likely to experience several changes in both the amount and source of his or her income. Along with these changes in income, there are likely to be changes in how much value the native places on money and possessions. These changes in income and values will not necessarily be all losses or all gains; they may be, but there may instead be some gains followed by sudden losses (or vice versa). Whether linear or not, these changes in income and values will all likely illuminate for the native different facets of their natal Pluto Sign placement. For example, someone with Pluto in Libra (whose Plutonic theme is empowerment through fairness, whether that fairness be in principle or in aesthetics) may find that an increase in earning capacity comes with an increased infatuation with the finer things in life at the expense of integrity, while a decrease in earning capacity forces them to discover that equality in relationships (and the integrity it takes to maintain this equality) is more important than flattering company or possessions. In this example, the Pluto in Libra native's increased earning capacity is accompanied by a more distorted sense of value, since sacrificing social fairness for things that look beautiful is what creates crisis for those with Pluto in Libra as Pluto transits the 2nd House. In contrast, decreased earning capacity may actually empower the native to get more of what they want out of life and leave a good legacy for those who come after them. This isn't to say that an increase in income always leads to a loss of integrity, just that while transiting Pluto is in the 2nd House, an apparent triumph in one area (income or values) may be accompanied by a setback in the other. As with anything else related to Pluto, integrity is the benchmark by which to measure and predict overall success or failure: If the native is doing more of what he or she always said he or she wanted to do while Pluto transits the 2nd House, then he or she is on the right track. If he or she is holding him- or herself back from life-long goals for the sake of immediate material success, then the native is probably setting him- or herself up for crisis and potential failure.

**3rd House**: Pluto's transit of the 3rd House indicates a time of re-evaluation and radical change in one's daily routine. The amount of time one spends travelling locally or communicating with others on a daily basis is likely to change drastically as Pluto enters and moves through this House. Much of this will probably come about because of the native questioning long-standing habits of thought and movement. He or she may begin to ask questions like "Why do I think I have to drive to the grocery store around the corner for a candy bar?

What would just walking there be like?" or "Do I really want to spend as much time as I do socializing online every day?" or "Why do I avoid talking to people on the phone?" Questions like these may (or may not) lead to significant changes in how the native conducts their daily life.

Pluto transiting the 3rd House has some other potentially interesting effects. It can indicate a time when the native's siblings come to play a much greater role in the native's daily life than they have previously. If it occurs this will probably be a positive thing, bringing brothers and sisters closer together for mutual support (although the initial events that bring them closer may be traumatic in some way). Another potential effect is on the native's self-education. The native may take up a new or expand on an already existing area of study, without the benefit of a formal education in this area. Any form of self-directed study is likely to prove very successful while Pluto transits this House, as the native learns new and important skills or information. These skills and information are likely to prove useful to the native in their daily life, whether or not they represent something the native can use on the job or as a career.

**4th House**: The home and one's family past is what is likely to be re-examined and undergo radical change as Pluto transits this House. The changes in everyday routine that likely occurred as Pluto transited the 3rd House are likely to truly "hit home" as Pluto transits the 4th. This can manifest as a change of residence (whether moving or major remodeling), or a change in one's beliefs or private behavior that have existed since childhood. There is a possibility of conflict with one's parents while Pluto is in this House—usually brought on by parental unwillingness to let go—or of the death or separation of the native's parents.

All of us hang on to childish behavior or beliefs in some form, treasuring them and carrying them with us throughout our lives. As Pluto transits a person's 4th House, it forces the native to examine these behaviors and beliefs in the objective light of adult consciousness. If these childhood relics are not serving the native (if they compromise the native's integrity), hanging on to them will probably create crisis in the native's life as Pluto moves through the 4th House. Therapy of some kind may be useful during this transit as a way to gain perspective regarding what is happening.

**5th House**: During Pluto's transit through the 5th House, choices the native makes regarding any 5th House-related phenomena (artistic or other "creative" work, children, love-affairs, attention-attracting behaviors or endeavors) are likely to have long-term consequences for the native. Obviously any artwork or other creative endeavor undertaken and completed during this time will likely have a significant impact on the native's life (whether or not it gains recognition by the general public), but creativity is much more than art. The vast majority of people express their creativity primarily through their children ("procreativity") and through their romantic or dating lives, and secondarily through their methods of attracting the attention of others. As Pluto moves through a native's 5th House, whatever experience the person creates with their children, their romantic contacts, their attention-getting behavior and any artistic or otherwise creative endeavors will likely be with them for the rest of their lives. The native may feel a subconscious sense of urgency in these "creative" areas, and along with the intense desire to make sure everything is perfect in these areas may come a tendency to over-control that will create some of the very crises that the native fears. Conflicts or other complications with one's own children (or even threats to their health) are likely to materialize while Pluto transits the 5th House. There may be one or more love affairs—even if the native is married—with a "fatal attraction" quality to them, where the native knows the partner or the relationship is no good for them but can't seem to pull themselves away. The native may attract the attention of dangerous people during this time, or put his or her life needlessly at risk in an attempt to do something memorable. These situations are not the workings of fate (although they may seem to be at the time), but rather the native subconsciously creating crisis for the sake of bringing attention to needs and desires that he or she has repressed so effectively they've become fully unconscious. Pluto does not allow these things to remain buried, but brings them to the surface of the native's life in a way that's impossible to avoid. The consequences of the choices the native makes in dealing with them will probably stay with the native for the rest of his or her life. Quick fixes and reflexive reactions will probably lead to further crisis, while taking the time and effort to cultivate consciousness—and thereby choosing with integrity—will likely lead to greater empowerment.

**6<sup>th</sup> House**: Pluto brings the native's health and employment into focus as it transits the 6<sup>th</sup> House. Long-term health issues that the native has long-ignored (such as poor diet) may reach a crisis point while Pluto transits the 6<sup>th</sup> House. If the native chose to put his or her life or health at risk as an attention-seeking device while Pluto was in the 5<sup>th</sup> House, that behavior could cause serious injury or death while Pluto transits the 6<sup>th</sup> House. Letting go of treasured but physically unhealthy behavior (and forming new healthy habits) will probably be called for while Pluto transits the 6<sup>th</sup> House. Although this may initially involve some pain and self-denial, if the native is willing to endure this, he or she is likely to ensure him- or herself much more resilient health and a longer life as a result.

Holding on to one's old job may be very difficult while Pluto transits the 6<sup>th</sup> House. Health problems or friction with superiors may cost the native a comfortable work position, and force him or her to cast about for not only a new job, but a better fit for his or her overall profession. If the native has been embracing this search and is being truly open to what circumstances are telling him or her, this new profession will likely appear in the native's life shortly before Pluto enters the 7<sup>th</sup> House.

**7<sup>th</sup> House**: More than in any other House, Pluto's effects in the 7<sup>th</sup> House are most likely to involve projections of the native's inner needs and "complexes" onto other people and external events. Mostly, this will manifest in the native's partnerships, in business, marriage and even "open enemies" (those who have an acknowledged hostility to the native and actively work against him or her). Usually these effects will manifest within existing partnerships rather than creating new ones, but if a new partnership forms during this transit (and survives Pluto's stay in the 7<sup>th</sup> House) it will probably last the rest of the native's life. The native is likely to discover compulsions and needs in their partnerships that were previously hidden from them, and possibly feel compelled to try to continue or expand their partnership through dishonest means (or experience a partner trying gain advantage in the partnership dishonestly). For better or worse, this will necessitate the transformation of the kind of partnership the native has with the other person; this is unlikely to bring an end to a partnership (unless one partner or the other dies), and more likely to lead to a transformation of the partnership. This could take a negative form (a marriage partner becomes an open enemy) or a positive form (the native may partner successfully with a former enemy, such as a business rival, to achieve a shared objective). However these transformations manifest, they are likely to greatly shape how the native approaches and experiences similar partnerships with others for the rest of his or her life.

As with Pluto transiting the 4th House, psychotherapy may be very useful for the native as Pluto transits the 7th House for the purpose of gaining some perspective. Disturbing behaviors the native sees in others or disruptive effects partners have on the native while Pluto transits the 7th House are likely to be things that the native exhibits him or herself, although the native may have been totally unconscious of these behaviors (or their effects on others) before this Pluto transit. Trying to treat any problems that come up in partnerships while Pluto transits the 7th House purely by "fixing" external circumstances will probably just extend or deepen the crisis. If the native has the integrity to see how they are practicing the same kind of disturbing behavior immediately visible in their partners, he or she will more likely find resolution and real empowerment by the time Pluto leaves the 7th House.

**8th House**: Any dysfunctional partnerships that survived Pluto's journey through the 7th House are likely to reach a final crisis while Pluto is in the 8th. Whether this means divorce, bankruptcy or some other permanent severing of ties, partnerships that are based on dishonesty or coercion are likely to end definitively while Pluto transits the 8th House. Although it's possible for someone to come away from these circumstances in a better financial position than they were in before, the native with transiting Pluto in their 8th House will more likely lose than win in these situations.

This transit is very likely to reveal where the native has developed unhealthy dependencies. These dependencies could be based on other people, institutions or even ideas, but whatever the nature of these dependencies they are likely to cost the native greatly both financially and emotionally as Pluto transits the 8th House. Recognizing and detaching from these dependencies quickly will probably be key. Not only will it prove less expensive for the native the more quickly the native walks away, but the native will then be able to move forward with far less baggage and so be more effective in reaching their personal goals.

**9th House**: This is one of the gentler Pluto transits. Any crises that occurred while Pluto transited the 8th House are likely to leave the native questioning the validity of their world-view, and the native will probably have experiences and be exposed to ideas while Pluto is in the 9th House that give them a more complete understanding of the world and the native's role in it. These experiences and insights are likely to be educational without being overly traumatic, changing the native into a noticeably better person.

There are two behaviors that are a danger to the native during this transit, however: Crusading and obsession. If the native tries to cram their newfound wisdom down the throats of everyone they meet, then strained relationships—up to and including legal troubles—are a real possibility while Pluto transits the 9th House. Any relationships strained by the native's proselytizing will create crises that will stunt the native rather than help him or her to grow. The other danger is obsession; the native may become so wrapped up in their initial new experiences and perspectives, that they blind themselves to the even bigger revelations and more important experiences that follow on the heels of the first. The root of both these behaviors lies in the native taking him- or herself too seriously. Having the integrity to live and let live—and so allow others to freely experience the world as the native is now doing—is what will usually avoid or resolve crisis for those with Pluto transiting their 9th House.

**10th House**: Often, it seems that nothing less than immediate, total success will do for those with Pluto transiting their 10th House. The discoveries people make about themselves and the world while Pluto transits their 9th House may cause them to change direction in their careers. Sometimes, Pluto-in-9th-House experiences lead people to sacrifice integrity or the respect of others for quicker or more complete results while Pluto transits their 10th. They may believe that there is no time to waste, that they must do what they now know needs to be done as quickly as possible to make up for lost time (or past sins) and to ensure they don't miss their window of opportunity.

Pluto does drive one toward total success while it transits the 10th House, but changing careers and using tactics where the ends justify the means are indications that the native still doesn't really understand what their real purpose is in this life, no matter how certain of "the truth" he or she may feel. However rushed the native may feel, he or she really does have the luxury of time under this transit. Having the integrity to be patient with oneself—and so gain patience with others—while one's true life-calling reveals itself, is likely to be very important as Pluto transits a person's 10th House. Patience will let the native's true life-calling unfold at its own pace and with the willing cooperation of everyone who needs to be involved. This will help the native avoid or mitigate crises that arise while Pluto transits his or her 10th House.

**11th House**: Friendship will trump acquaintanceship as Pluto moves through a person's 11th House. Acquaintances the native may have kept in his or her life simply because their company was pleasant or convenient (but lacking any emotional bond) are likely to fall by the wayside as Pluto moves through the native's 11th House. These acquaintanceships are likely to be replaced by real friendships, in the form of a few people (or a distinct group or organization) with whom the native is able to form lasting emotional connections with.

Generally this effect is a very good thing, but there are a couple of things to watch out for. Any new friendships are best based on emotional experiences rather than material gain; some people find it tempting under this transit to use the new, deeper relationships they now develop with others as a means to a material end. Doing this is a mistake. The other thing to watch out for is an "ends justify the means" mentality when pursuing one's long-term goals. This is somewhat similar to Pluto's effect in the 10th House, but in the 11th House the native may be tempted to join groups that pursue extreme or otherwise unsavory tactics to attain desired ends, rather than the native taking the initiative him or herself. However, the means always influence the ends that are being worked toward, and using injustice to work toward a just goal only ensures that the reality of the goal remains unobtainable, even if the form of the goal is brought into being. "Peace through superior firepower" may indeed entail an end to hostilities, but it is a peace that is neither nourishing nor sustainable. The means always influence the ends; if amoral means are used to attain a moral goal, the goal will never really be attained.

The real basis of these new relationships is the native's changing sense of long-term goals. The ideals or goals the native has been working toward (or achieved) may prove inadequate or even childish as Pluto transits the native's 11th House, and as the native casts about for a new and better direction (in part by insisting on more authentic emotional experiences when interacting with others), new and better relationships will appear in the native's life. These new relationships are likely to play a large part in shaping what long-term direction the native will move in. Although the native will probably learn much from the new associations formed while Pluto moves through his or her 11th House, these associations are likely to both begin and remain as friendships (in the sense of an interaction between equals) rather than any kind of teacher-student dynamic.

**12th House**: Solitude is what will likely empower the native initially as Pluto transits his or her 12th House, while self-defeating behavior rooted in guilt often creates crisis during this time. As with Pluto transits to a native's 4th and 7th Houses, this transit represents a time when some form of psychotherapy would be particularly helpful for the native. Things the native has done or experiences the native has had for which they hold unacknowledged guilt are likely to lie at the root of any self-defeating behavior and opposition from others they experience while Pluto transits his or her 12th House. Acknowledging these actions, experiences and feelings, and addressing these things in a healthy way by forgiving or making amends where appropriate makes for a rapid rise out of crisis during this transit, empowering the native to express him or herself more fully as Pluto moves into the 1st House.


**Aspects From Transiting Pluto to Natal Planets**

Looking at Aspects from transiting Pluto to natal Planets can also be instructive. These Aspects can show what parts of the native's personality are involved with or most strongly affected by the transformation transiting Pluto brings. If positive, these Aspects can show what components of a person's life may be helping this transformation. If negative, these Aspects can show what may be blocking transformation, while natal Planets that don't Aspect transiting Pluto are probably not involved in the transformation transiting Pluto brings, even if they may seem to be. Looking up the relevant Aspects in the "Pluto In Aspect to Other Planets" section in Chapter 2 may be helpful. The difference between a natal and a transiting Aspect is straightforward: a transiting Aspect lasts for a relatively brief time, while a natal Aspect is in effect for the native's whole life. With Pluto, the effect described by a transiting Aspect to another Planet is likely to last for about a year or so.

While I've found transiting Pluto invaluable for predicting and understanding trends, I haven't found it useful for predicting specific events for groups or individuals.

## Chapter 5: Synastric Influence

For any readers who may not know, "synastry" is a sub-field of astrology that studies how placements in two people's birth-charts affect each other and thereby influence the relationship the two people may develop. Pluto's "synastric influence" is the influence it has on the relationships between people. Although Pluto's influence in synastry does not seem as significant as that of the personal Planets, it can show a few things worth keeping in mind. In a general sense, Pluto will show where power flows in a relationship; looking at what House one person's Pluto falls in the other person's chart will show where the Pluto person will have influence in the House person's life. I've also noticed that when one person's Planets Aspect another person's Pluto, it's usually the Pluto person that has the power in the relationship, unless the Aspecting Planet is one of the feminine Planets—Venus or the Moon—in which case the Venus or Moon person will tend to have power over the Pluto person.

I've observed that Aspects from one person's Pluto to another's don't often seem significant if the individuals have Pluto in the same Sign. A difference in two people's Pluto's Sign placement can be important, however, even if the two Pluto placements are not in orb to form an Aspect. If two people have Pluto in different Signs, they are of different generations (even if they are close in age), and so will have different ways of empowering themselves, different ways of creating crisis, and very different ways of interacting with groups. This can be a positive feature in the relationship if the partners are aware of this and know to give each other the space and different kinds of support each needs, but it can be a negative if one or both are unaware of these differences and expects the other to need and interact with groups as they do. A lack of awareness of or open-mindedness to this difference can lead to misunderstandings regarding why what empowers one person does not empower the other, or why crisis manifests for each of them in different ways and at different times.

# Afterword

All this new information on Pluto's influence raises a couple of important questions. Firstly, does the Pluto archetype—an underworld god of the dead and of money—really suit the effect the Planet appears to have on events here on Earth?

And if not, is there an archetype which seems more appropriate?

My answers to these questions are that the Pluto archetype is not the best fit for what I've observed of Pluto's influence, and there is indeed an archetype that seems much more appropriate to this Planet. The archetype that seems to fit best is Heracles (often known nowadays by the Romanized spelling of Hercules), the Greek hero.

I see many parallels between the Heracles myth and the Planet Pluto's influence. Pluto's influence creates the Hero's Journey for all of us. Heracles was considered by the ancient Greeks to be their greatest hero, setting the standard to which all other heroes were measured. Heracles was the only demi-god—half human and half divine being—that the Greeks widely recognized as worthy of worship, and so Heracles held the distinction of being the only Greek hero with his own religious following. This seems a close parallel to the confusion the celestial body called Pluto creates in the astronomical community, as it combines some typically Planet-like (and so "godlike") traits with some non-Planet-like (and so "mortal") traits. The fact that this celestial body continues to be recognized as a Planet by the general populace and given significance by the astrological community despite many astronomers not considering it a real Planet, seems to parallel the Greeks' unique worship of Heracles as a demi-god.

Heracles was famed for his strength, a direct parallel to power in a pre-industrial age, and power appears to be the essence of Pluto's influence. Heracles and Hermes (the god whom the Romans named Mercury) were the co-patrons of *gymnasia* and *palestrae* (wrestling schools), which corresponds well with the extremely high incidence of Pluto-Mercury Aspects in the charts of famous sports figures (see the Statistical Data Appendix).

Heracles was a very passionate individual, capable of both incredible loyalty (he once wrestled Thanatos, Death himself, for the sake of his friend Admetus) and implacable vengeance such as that against Augeas. The passion, loyalty and sense of vengeance Heracles displayed all align closely with the Planet Pluto's observable influence, especially on the Sign of Scorpio.

Heracles was constantly getting himself into trouble through abusing his great strength—his power—forcing him to spend much of his life making amends. The feats Heracles is best remembered for— his Twelve Labors—were undertaken as a way to make amends for killing his wife and children in a fit of irrational temper. This can be seen as a parallel to the empowerment-crisis-empowerment process the Planet's influence seems to symbolize for us individually and collectively here on Earth.

Pluto's curious astrological synergy with the Planet Uranus could also be explained by viewing the Planet Pluto as Heracles, in combination with (as first proposed by Richard Tarnas) viewing the Planet we call Uranus as Prometheus. In Greek mythology, as punishment for his iconoclastic non-conformity—classic traits of the Planet Uranus' astrological influence—in defying Zeus to give mankind fire, the Titan Prometheus was chained to a rock, condemned to this confinement and to having his liver eaten from his living chest, every day, for eternity. Zeus eventually relented, but he did not immediately set Prometheus free. Instead, Zeus had Heracles broker a deal between Chiron (an immortal centaur with an un-healable wound) and Prometheus, where Chiron would be allowed to die in Prometheus' place and so set Prometheus free, allowing the Titan his freedom as long as he did not abuse it by once again disobeying Zeus.

Although the Planet Uranus (and its arguably more appropriate archetype Prometheus) usually represents the triumph of non-conformity, when interacting with Pluto via Aspect this Planet's fundamental influence is uniquely reversed. Looking at Pluto (Heracles) as an influence that brings the rebel Uranus (Prometheus) to see the benefits of conformity seems to provide a mythological parallel to the observable astrological interaction of these two Planets.

Finally, the goddess Hera—Heracles' step-mother—had tremendous influence over Heracles' life, and almost always used this influence negatively. Yet, upon the death of Heracles' mortal being, Hera allowed Zeus to grant Heracles' inclusion into the Greek pantheon. None of Zeus' many other illegitimate offspring ever accomplished this ultimate ascension. Hera made an exception for Heracles—and so made Heracles' greatest feat, that of becoming a god, possible—because over the course of his mortal life, Heracles had never stopped humbly begging her forgiveness (his very name was an attempt to placate her). Rather than becoming bitter and vengeful, his integrity and humility toward the gods and toward mortals proved his worth to both Zeus and Hera in a way his sheer strength never could have.

This seems to form an excellent parallel with the Planet under discussion on more than one level. Not only does this myth jibe with how the feminine Planets (and especially the Moon, symbolizing the mother) influence Pluto in synastry, but also with how using integrity to overcome Plutonic crisis transforms the crisis into a boon, turning mortal (or immortal) enemies into friends.

Taking into account all of these correspondences, I believe that the Planet under discussion has been misnamed, and that it should bear Heracles' name. I also think it likely that the name and symbolism of Pluto is poisoning this Planet's influence with the darkness and fear accorded an underworld deity. Using what observation shows to be this Planet's proper name and symbolism— Heracles—would allow it to be recognized as the humanizing and heroic influence in our lives that it actually is. As he did with his step-mother, Heracles is begging us to forgive him for not meeting our expectations, and to accept his noble and immortal nature for what it is. If a petty and conniving shrike like Hera finds it in her heart to get over herself and love him, can we who think ourselves more emotionally mature than her do any less?

I have used the name Pluto throughout this book when referring to the Planet under discussion, rather than using what I see as its proper name of Heracles, for the sake of clarity. I hope that as more people are exposed to the research I've done and the conclusions I've put forth regarding this Planet's influence, more and more people will come to recognize the wisdom of renaming this Planet and allow its proper correspondences to fully infuse our consciousness.

In closing, I don't view Pluto as being any more or less significant than any of the other Planets. It is a major gem in the astrological jewel that is our Solar System, and however the astronomical debate is eventually settled, I believe Pluto has proven to have an equal astrological significance to those other physical bodies we call Planets. Although certainly powerful enough to stand on its own, I believe that only in the context of the other Planets' influences—the individual consciousness and choice represented by the Sun, our emotional and family histories as described by the Moon, the different ways we perceive and communicate about our world as shown by Mercury, and so on—can we fully understand and make the most of Pluto's influence.

## A Final Note

The influences and correspondences described in this book have been of a uniformly pragmatic nature, other than those mentioned in the Afterword. These pragmatic correspondences can be applied and studied regardless of whether or not Pluto is considered to be a Planet, or whatever name is assigned to this particular celestial body. Since it seems that there are an awful lot of these immediately useful correspondences, and that they have been missing from astrological discourse, I thought it best to bring these widely useful discoveries to light as soon as possible. Hence, this book has been dedicated specifically to those correspondences that are independent of the celestial body's perceived identity.

There are, however, other more abstruse—but equally fascinating—influences this Planet appears to have on Earthly phenomena, which have not been detailed in this book. These correspondences seem to be occulted by the filter of Pluto's myth, as they only become discernible when this Planet's influence is viewed through the lens of the Heracles myth. These abstract correlations are interesting and complex enough that they deserve a book of their own. I plan on releasing a book describing these other correlations shortly. It is my hope that with the release of the current book, the astrological community will come to see the wisdom of renaming the Planet we now generally call Pluto. This will pave the way toward the further understanding made possible by the more abstract correspondences I will be detailing in my next book.

# Appendix: Statistical Data

The data contained in this Appendix is the raw material from which I have obtained many of the inferences and conclusions I have drawn throughout the book. I have included this data here in as complete a manner as possible, so that interested readers may follow my research for themselves if they so wish. In any serious field of study, it's important to let one's peers review one's basic research, both the undigested data and the process used to reach conclusions, so that one's professional peers may refine these things in a constructive way. This statistical appendix is my attempt at fostering this openness and refinement, so that my research in particular—and the field of astrology in general—will benefit.

Also, there are some conclusions and patterns that show up in the statistical analysis of Pluto's influence that I did not describe in detail in the text of this book. The secondary purpose for this appendix is to present these conclusions and patterns in detail.

### Baseline

To understand the exceptional, there has to be a norm to give it context and to measure it against. In an attempt to find out what is "normal" for Pluto Aspects and make my baseline group's distribution of Pluto Aspects as representative of the general population as I could, I used as a statistical sample of the general population a group of 20 natal charts, 18 of which I've obtained from clients over the years and two of which belong to accomplished public figures (Alan Greenspan and Temple Grandin). I included two famous and successful figures because if one looks at any population as a whole, a minority does always find itself at the edge of the bell-curve, and I felt it was important to have this minority represented to form a proper representation of the general population. Other than the two famous names I've just mentioned, I'm going to withhold the names connected with the other charts I've used for my baseline in the interests of privacy. I will, however, give the birth dates, times and locations associated with each of the other 18 charts that I've used in chronological order, as well as for the charts of Mr. Greenspan and Ms. Grandin. This should provide any interested astrological

103

researcher with all the data they need to reproduce or critique my sample group, while preserving the privacy of those who are not already in the public eye.

| Name | Birth Date | Birth Time | Birth Location |
|------|-----------|-----------|----------------|
| Alan Greenspan | 3-6-1926 | noon | New York NY |
| -- | 4-17-1946 | 10 AM | Trenton NJ |
| Temple Grandin | 8-29-1947 | 2:30 PM | Boston MA |
| -- | 5-20-1955 | 2:16 PM | Gothenburg Sweden |
| -- | 6-23-1955 | 2:00 PM | South Bend IN |
| -- | 11-12-1955 | 4:34 PM | Newark NJ |
| -- | 12-10-1957 | 5:19 AM | Richland Center WI |
| -- | 4-10-1958 | 10:30 AM | Columbia MO |
| -- | 7-27-1963 | 5:52 PM | South Charleston SC |
| -- | 4-30-1964 | 5:45 PM | Denver CO |
| -- | 7-1-1964 | 1:37 PM | Honolulu HI |
| -- | 6-26-1965 | 6:30 AM | Detroit MI |
| -- | 3-12-1970 | 10:03 PM | Macon GA |
| -- | 12-4-1970 | 4:12 PM | Richmond VA |
| -- | 1-28-1975 | 4:32 AM | Detroit MI |
| -- | 5-6-1981 | noon | Greeley CO |
| -- | 6-17-1981 | 9:17 PM | Denver CO |
| -- | 12-15-1987 | 11:07 PM | Laramie WY |
| -- | 10-24-1998 | 9:30 AM | Aurora CO |
| -- | 2-12-2009 | 6:05 AM | Wheatridge CO |

From these 20 charts I formed a baseline for the percentages of Pluto's Aspects to each of the other Planets. To obtain these percentages, I took the total number of charts (20), and divided the number of charts that had a Pluto Aspect to the Planet in question by that total number. For example, out of the 20 total charts examined, 8 had a natal Aspect between Pluto and the Sun. 8 divided by 20 yields .4, which is equivalent to 40%. Below is the complete baseline percentage data:

| Aspect | Frequency |
|---|---|
| Pluto-Sun | 40% |
| Pluto-Moon | 45% |
| Pluto-Mercury | 25% |
| Pluto-Venus | 30% |
| Pluto-Mars | 25% |
| Pluto-Jupiter | 20% |
| Pluto-Saturn | 40% |
| Pluto-Uranus | 25% |
| Pluto-Neptune | 90% |

Given the degrees of orb I generally assign to major natal Aspects—10 degrees to each possible conjunction, 5 degrees to each possible sextile, 8 degrees to each possible square, 6 degrees to each possible trine, 1 degree to each possible quincunx, and 9 degrees to each possible opposition for a total of 59 possible degrees out of the 360 contained in the circle of the Zodiac—pure probability indicates that an Aspect should show up between any two Planets about seventeen percent of the time, or in one or two charts out of every ten. As can be seen from the baseline group, even given a generous margin for statistical error this is not always the case when Pluto Aspects other Planets. Although four of the nine Planets do fall within ten percentage points of the indicated statistical likelihood of 17% when Aspecting Pluto in the baseline group, four Planets Aspect Pluto significantly more frequently than chance would indicate (The Sun, the Moon, Venus and Saturn), and a fifth Planet almost always Aspects Pluto in this statistical sample (Neptune).

### Other Sample Groups

After establishing a baseline, I felt that sample groups of more specialized charts—such as those of the rich and famous—could now be compared and usefully analyzed. I've broken down my other sample groups by how society has categorized the contributions of each group's members.

I've put together 10 specialized sample groups and listed them separately below. Each sample group I've put together contains twenty members. To represent each sample group I've provided a list of the major figures in the areas of Astrology, Business,

Entertainment, Infamous, Literature, Politics, Royalty, Science, Sports, and The Arts, and provided a separate table for each group showing what percentage of charts in the group had Aspects to Pluto, by Planet.

Every sample group I've put together has provided some very interesting and often very distinct variations from the baseline group's distribution of Pluto placements. Following The Arts sample group data is a chart that summarizes these variations.

## Astrologers

**Name**
Alice Bailey
Donna Cunningham
Jeanne Dixon
Cyril Fagan
Steven Forrest
Linda Goodman
Jeff Green
Robert Hand
Karl Ernst Krafft
William Lilly
Marion March
Joan McEvers
Nostradamus
Robert Pelletier
Dane Rudhyar
Howard Sasportas
Ebenezer Sibly
Richard Tarnas
Noel Tyl
Wilhelm Theodor H.
Wulff

| Aspect | Frequency |
|---|---|
| Pluto-Sun | 20% |
| Pluto-Moon | 30% |
| Pluto-Mercury | 10% |
| Pluto-Venus | 30% |
| Pluto-Mars | 35% |
| Pluto-Jupiter | 35% |
| Pluto-Saturn | 45% |
| Pluto-Uranus | 35% |
| Pluto-Neptune | 50% |

## Businesspeople

**Name**
Neil Bogart
Richard Branson
Warren Buffett
Suzanne DePasse
Walt Disney
Elena Ford
Henry Ford
Bill Gates
John Gotch
Sir Hugh Greene
Hugh Hefner
Leona Helmsley
L. Ron Hubbard
Lee Iacocca
Ferruccio Lamborghini
Rupert Murdoch
Ross Perot
John D. Rockefeller
Donald Trump
Ted Turner

| Aspect | Frequency |
|---|---|
| Pluto-Sun | 50% |
| Pluto-Moon | 25% |
| Pluto-Mercury | 20% |
| Pluto-Venus | 25% |
| Pluto-Mars | 25% |
| Pluto-Jupiter | 15% |
| Pluto-Saturn | 30% |
| Pluto-Uranus | 35% |
| Pluto-Neptune | 30% |

## Entertainment

**Name**
Paula Abdul
Woody Allen
Joan Baez
Lucille Ball
Antonio Banderas
Drew Barrymore
John Belushi
Frank Sinatra
Steven Spielberg
Cat Stevens
Lily Tomlin
John Travolta
Tina Turner
Tracey Ullman
Eddie Vedder
Denzel Washington
Mae West
Oprah Winfrey
"Weird Al" Yankovic
Neil Young

| Aspect | Frequency |
| --- | --- |
| Pluto-Sun | 20% |
| Pluto-Moon | 40% |
| Pluto-Mercury | 45% |
| Pluto-Venus | 35% |
| Pluto-Mars | 45% |
| Pluto-Jupiter | 45% |
| Pluto-Saturn | 40% |
| Pluto-Uranus | 15% |
| Pluto-Neptune | 70% |

## Infamous

**Name**
Moses Annenberg
Marshall Applewhite
George Blake
Ian Brady
Ted Bundy
Mark David Chapman
Jeffrey L. Dahmer
James Degorski
Lynndie England
Heidi Fleiss
John Wayne Gacy
Edward Gein
Mata Hari
J. Edgar Hoover
Jim Jones
Don King
Rodney King
David Koresh
Charles Manson
Lee Harvey Oswald

| Aspect | Frequency |
| --- | --- |
| Pluto-Sun | 40% |
| Pluto-Moon | 40% |
| Pluto-Mercury | 20% |
| Pluto-Venus | 30% |
| Pluto-Mars | 30% |
| Pluto-Jupiter | 40% |
| Pluto-Saturn | 50% |
| Pluto-Uranus | 40% |
| Pluto-Neptune | 35% |

# Literature

**Name**
Louisa May Alcott
Hans Christian
Andersen
Maya Angelou
Honore de Balzac
Charles Baudelaire
Anthony Burgess
Lord Byron
Albert Camus
Truman Capote
Lewis Carroll
Samuel Taylor
Coleridge
Michael Crichton
Charles Dickens
Emily Dickinson
T.S. Eliot
William Faulkner
Upton Sinclair
J.R.R. Tolkien
Jules Verne
Oscar Wilde

| Aspect | Frequency |
| --- | --- |
| Pluto-Sun | 45% |
| Pluto-Moon | 45% |
| Pluto-Mercury | 25% |
| Pluto-Venus | 25% |
| Pluto-Mars | 50% |
| Pluto-Jupiter | 45% |
| Pluto-Saturn | 10% |
| Pluto-Uranus | 25% |
| Pluto-Neptune | 35% |

## Politics

**Name**
Samuel Adams
Otto von Bismarck
Winston Churchill
Oliver Cromwell
Adolf Hitler
Herbert Hoover
Saddam Hussein
Andrew Jackson
John F. Kennedy Jr.
Martin Luther King
Henry Kissinger
Helmut Kohl
Joseph McCarthy
Richard Nixon
Cecil Rhodes
Maximilien
Robespierre
Franklin Delano
Roosevelt
Stalin
Margaret Thatcher
George Washington

| Aspect | Frequency |
|---|---|
| Pluto-Sun | 25% |
| Pluto-Moon | 20% |
| Pluto-Mercury | 35% |
| Pluto-Venus | 35% |
| Pluto-Mars | 35% |
| Pluto-Jupiter | 45% |
| Pluto-Saturn | 25% |
| Pluto-Uranus | 25% |
| Pluto-Neptune | 30% |

# Royalty

**Name**
Princess Aiko
Emperor Akihito
Albert II, Prince of
Monaco
Albert Victor, Prince
of Wales
Charles, Prince of
Wales
 Dianna, Princess of
Wales
Marie-Antoinette
Mary, Princess Royal
Queen Mary
Maud, Queen of
Norway
Maximilian I, Holy
Roman Emperor
Michael, Prince of
Kent
Zenouska Mouwatt
Prince Naruhito
Marguerite of Navarre
Camilla Parker-
Bowles
Peter Phillips
Zara Phillips
Renee, Princess of
France
Sarah, Lady Chatto

| Aspect | Frequency |
| --- | --- |
| Pluto-Sun | 30% |
| Pluto-Moon | 40% |

| | |
|---|---|
| Pluto-Mercury | 30% |
| Pluto-Venus | 15% |
| Pluto-Mars | 35% |
| Pluto-Jupiter | 45% |
| Pluto-Saturn | 35% |
| Pluto-Uranus | 30% |
| Pluto-Neptune | 65% |

## Science

**Name**
Alexander Graham
Bell
Louis Braille
Nicholas Culpepper
Rene Descartes
Thomas Edison
Albert Einstein
Havelock Ellis
Michael Faraday
Galileo
Robert Hooke
Edwin Hubble
Christiaan Huygens
Johannes Kepler
James Lovelock
Nevil Maskelyne
Isaac Newton
Louis Pasteur
Thomas Say
Nikola Tesla
Karl Weierstrass

| Aspect | Frequency |
|---|---|
| Pluto-Sun | 50% |
| Pluto-Moon | 15% |
| Pluto-Mercury | 20% |
| Pluto-Venus | 25% |
| Pluto-Mars | 45% |
| Pluto-Jupiter | 35% |
| Pluto-Saturn | 40% |
| Pluto-Uranus | 25% |
| Pluto-Neptune | 55% |

## Sports

**Name**

Muhammed Ali
Kenny Anderson
Arthur Ashe  PM
Vicki Aragon Baze
David Beckham
Peter Blake
Bjorn Borg
Peter Brock
Howard "Hopalong"
Cassady
Wilt Chamberlain
Michael Chang
Larry Christiansen
Jimmy Connors
Mary Decker
Jean Driscoll
Chris Evert
Juan Miguel Fangio
Rollie Fingers
Dawn Fraser
Rudy Galindo

| Aspect | Frequency |
|---|---|
| Pluto-Sun | 20% |
| Pluto-Moon | 35% |
| Pluto-Mercury | 60% |
| Pluto-Venus | 20% |
| Pluto-Mars | 60% |
| Pluto-Jupiter | 40% |
| Pluto-Saturn | 35% |
| Pluto-Uranus | 35% |
| Pluto-Neptune | 75% |

## The Arts

**Name**
Diane Arbus
Steve Artley
Jean-Baptiste
Carpeaux
Rick Castro
Paul Cezanne
Salvador Dali
Albrecht Durer
Mariano Fortuny
Albert Goodwin
Keith Haring
Gustav Klimt
Henri Matisse
Michelangelo
Pablo Picasso
Cindy Sherman
Henri Toulouse-
Latrec
Vincent Van Gogh
Johannes Vermeer
Leonardo da Vinci
Andy Warhol

| Aspect | Frequency |
| --- | --- |
| Pluto-Sun | 25% |
| Pluto-Moon | 20% |
| Pluto-Mercury | 30% |
| Pluto-Venus | 30% |
| Pluto-Mars | 40% |
| Pluto-Jupiter | 35% |
| Pluto-Saturn | 20% |
| Pluto-Uranus | 25% |
| Pluto-Neptune | 40% |

I've provided a chart, below, that shows the significant deviations from the baseline group that each of the other statistical groups display in the frequency of their Pluto Aspects. A "▲" indicates that a particular group (represented by the column) has an Aspect to a given Planet significantly more frequently than the baseline group, while a "▼" indicates that a particular group has an Aspect to a given Planet significantly less frequently than the baseline group. When comparing the frequency of Pluto Aspects in groups of famous people to the frequency of Pluto Aspects in the general population, I've ignored differences of ten percentage points or less from the baseline. Since I'm using such small sample groups, I view differences of ten percent or less as likely being statistical noise, and so have focused my attention on those with a frequency difference greater than 10 percent.

| | Ast | Biz | Ent | Inf | Lit | Pol | Roy | Sci | Spo | Art |
|---|---|---|---|---|---|---|---|---|---|---|
| Sun | ▼ | | ▼ | | | ▼ | | | ▼ | ▼ |
| Moon | ▼ | ▼ | | | | ▼ | | ▼ | | ▼ |
| Mercury | ▼ | | ▲ | | | | | | ▲ | |
| Venus | | | | | | | ▼ | | | |
| Mars | | | ▲ | | ▲ | | | ▲ | ▲ | ▲ |
| Jupiter | ▲ | | ▲ | ▲ | ▲ | ▲ | ▲ | ▲ | ▲ | ▲ |
| Saturn | | | | | ▼ | ▼ | | | | ▼ |
| Uranus | | | | ▲ | | | | | | |
| Neptune | ▼ | ▼ | ▼ | ▼ | ▼ | ▼ | ▼ | ▼ | ▼ | ▼ |

There are several interesting facts that are readily apparent from this chart.

Each sample group appears to have a distinct "profile" of Pluto Aspect frequency that separates it not only from the baseline group, but also from the other sample groups. This seems to imply the possibility that Pluto has a distinct influence on (or at least a distinct correspondence with) different areas of human activity in combination with other Planets, but more research is needed before much more can be said with certainty on this topic. I think it's interesting to note that these profiles appear to show patterns, not only of the increased incidence, but also of the increased *absence* of Pluto's influence relative to the baseline population, in the specific areas of human activity represented by each of the sample groups.

It also appears that Aspects from Pluto to Mars and from Pluto to Jupiter are found much more frequently in the charts of those who become famous than either probability or their distribution in the baseline population would indicate. In addition, it appears that Aspects from Pluto to all five of the Planets that are Aspected more often than chance would indicate in the baseline group—the Sun, the Moon, Venus, Saturn and Neptune—are significantly less common in the charts of those who are famous. This seems to be least true of Venus, and most true of Neptune.

Only three of the ten groups analyzed here (Entertainers, the Infamous and Sports figures) have profiles that are weighted toward a higher incidence of Pluto's Aspects rather than their absence. Another two groups—Literary figures and those in Science—appear to have neutrally weighted profiles regarding the presence and absence of Pluto's Aspects to other Planets, while the remaining five groups actually seem to be weighted towards the absence of Pluto's Aspects to other Planets relative to the baseline population.

The specific interpretations of these statistical trends I've made can be found in the Individual Influence chapter. See the general description of each Planet's Aspect to Pluto (Sun-Pluto, Moon-Pluto, etc.).

The two final things to remember about the statistics I've compiled and collated here are that they are a beginning—not a completion—of the possible further research concerning Pluto's observable influence, and that like any statistics, the trends implied by the chart above are generalizations, and are not absolute rules. I found no group that always had a Pluto Aspect to a particular Planet, nor any group that was completely without Aspects from Pluto to a particular Planet. There are famous people with a Pluto-Neptune Aspect in their natal chart, and not everyone with a natal Pluto-Jupiter Aspect is rich and famous.

My observation is that the choices each of us make are a more powerful factor when determining the course of our lives than any astrological influence. Like any other astrological influence, Pluto only inclines, it does not compel. I believe it is only because we have remained largely ignorant of Pluto and how it is influencing us that its effects often appear to be the mysterious workings of fate.

www.ingramcontent.com/pod-product-compliance
Lightning Source LLC
Chambersburg PA
CBHW060545100426
42742CB00013B/2464